TELECOM SERVICE ROLLOUTS

A Best Practices Guide

RAJOO NAGAR

D1450538

McGraw-Hill
New York   Chicago   San Francisco   Lisbon
London   Madrid   Mexico City   Milan   New Delhi
San Juan   Seoul   Singapore   Sydney   Toronto

McGraw-Hill

A Division of The McGraw-Hill Companies

1 2 3 4 5 6 7 8 9 0 AGM/AGM 0 8 7 6 5 4 3 2

ISBN 0-07-139112-6

The sponsoring editor for this book was Marjorie Spencer and the production supervisor was Pamela A. Pelton. It was set in Fairfield Medium by Patricia Wallenburg.

Printed and bound by Quebecor/Martinsburg.

McGraw-Hill books are available at special quantity discounts to use as premiums and sales promotions, or for use in corporate training programs. For more information, please write to the Director of Special Sales, McGraw-Hill Professional, Two Penn Plaza, New York, NY 10121-2298. Or contact your local bookstore.

 This book is printed on recycled, acid-free paper containing a minimum of 50 percent recycled, de-inked fiber.

To my mother and Simi

McGraw-Hill Telecommunications

CONTENTS

PREFACE

The writing of this book came about in an entirely unexpected manner. Prior to starting this project, I worked at a company that developed and sold high-end network infrastructure gear to competitive local exchange carriers. As the company's customer base dried up, I saw at close range the collapse of the telecom service provider as it began. But I had also seen (in even earlier days) the carrier emerge triumphant from the shackles of oligopolistic telecom practices—and these shackles were broken in part by technological advances and in part by the now (in)famous Telecom Act of 1996. As the industry saw the rise and fall of young and aspiring competitive carriers, an intense debate followed each collapse that occurred, and many questions were raised about the phenomenon that was taking place: Was this phenomenon entirely new or had it been duplicated before in some other industry? What resemblance, if any, did it bear to the dotcom bubble? What would happen next? And how would consumers be impacted? Already under contract to write a book on deploying broadband services in the multi-tenant unit (MTU) market, I started to research this topic in earnest, to the extent that it soon became in my eyes a much more compelling story than the one I was supposed to write. Midway through the MTU book project, I abandoned it and wrote this book instead.

Aside from the curiosity and the desire to document what happened, I had other reasons for writing this book. It is important to note that, although the first chapter focuses on the telecom failures primarily to expose mistakes and lapses in judgment, the rest of the book focuses on successful market strategies and best practices for deploying telecommunications services. And that is the main reason for writing this book—to research, uncover, and lay forth sound business and market strategies essential for succeeding in this volatile and shifting telecom environment. I believe that this is the first book that deals with the practical and business considerations of rolling out a telecommunications service. It is also the first book that explores the failures in telecom, discusses wrong turns taken by management, and suggests best practices that, if followed, may have kept these companies alive. Still, this book would perhaps never have been written had it not been for the persistence and guidance of Marjorie Spencer, senior telecommunications editor at McGraw Hill, who convinced me that indeed this was a topic that would be of immense benefit to the reader, and offered valuable insights throughout the writing of the book.

Once begun, the project took me much longer to complete than I had anticipated. Researching the telecom boom and bust and studying the market and its players with the intent of uncovering not only causes for failures and successes but precious nuggets of wisdom and best practices required tremendous diligence, many questionnaires, and phone interviews, and many months of sifting through hundreds of reports and research data that resided in many different locations—at research and analyst agencies, telecom publications, Wall Street and venture capital firms, and the telephone and equipment companies themselves. There were a couple of instances when I picked up the phone to call a telecom service provider and got a voice message at the other end that asked me not to attempt to contact employees, and to follow certain instructions if I was a supplier or distributor: it was obvious that the company was in the process of shutting down. At other times, I read about a service provider's great fundamentals in a Wall Street analyst's research report and some few weeks later the company declared bankruptcy—I then discovered through follow-on research that the company's chances of survival had been slim all along. And so it went, all the while adding to my repertoire of knowledge and insights about this industry.

Having many years of experience with the telecom community, I was curious to dig deeper into the fallen companies to see if I could discover any common threads among them that might have contributed to their downfall. This then led to my studying the companies that were managing to survive the storm, and see if they in turn shared common characteristics. Not surprisingly, they did. From putting a management team in place, to planning a market entry strategy, to rolling out a service, to servicing a customer base—telecom survivors over the past two years have shared a set of core values and fundamental business sense that have served them well. Their best practices are captured in this book as well.

Finally, there are several key things whose existence is fundamental to the success of the telecom industry and the players in it, and this book addresses the importance of these topics specifically as they relate to the delivery of broadband services:

- *A strong competitive landscape.* Competition is essential if consumers and businesses are to receive new and innovative services at competitive prices. The right to choose will only prevail if the local phone markets do not become monopolies (again) and end-customers are given the flexibility to choose their service provider.
- *The proliferation of applications.* Application support by independent software vendors is essential to end-customer acceptance of new telecom technologies. Price and platform stability can only be brought about by increased usage, which in turn is spurred by the availability of applications. Service providers must work aggressively to partner with application providers to bring enhanced services to the office and home.
- *Support for standards.* Proprietary protocols have given some service providers an unfair edge and blocked the introduction of new services. Industry-wide support for standard protocols both in the last mile and the backbone network is essential for interoperability between equipment and services, new service support, and open service creation.

In summary, the task of researching hundreds of articles and reports and talking to numerous telephony experts about my views and soliciting feedback was a process that took well over a year, but resulted in the book that follows. Some of the fundamental questions that I have attempted to answer in the chapters that follow include: What is the secret to survival in this surprisingly volatile industry? What sets the survivors apart from the failures? Is it operational efficiencies, or sound business models, or plain good timing? Does the answer lie in the use of leading-edge equipment and networks, or in amassing a huge customer base, or in long-range strategic planning? What can we learn from the companies who have survived these turbulent times and will live to play another round of the game? As the reader will discover, the answers reveal a great deal about the tough economic environment we find ourselves in, the changing habits of the end-user, and the newly restored criteria for long-term success in the telecom industry.

ACKNOWLEDGMENTS

The idea of writing this book came from Marjorie Spencer, senior telecommunications editor at McGraw Hill, so I owe a special thanks to her. Additionally, this book was made possible by the insight, support, and encouragement of several people, and was substantially influenced by my continuing interactions with a number of experts in this field. In particular, I would like to express my gratitude and appreciation to Christine Heckart, president of TeleChoice, for her valuable insights that helped build knowledge and provide perspective to this book. Most notably in Chapters 1 and 2, and in various places throughout the book, Christine provided tremendous insight into what it takes to be a successful telecommunications company. I would also like to thank the rest of the team at TeleChoice, including Tom Jenkins, Russ McGuire, and Pat Hurley, who provided valuable proprietary research material for inclusion in some of the chapters. Finally, at TeleChoice, many thanks to Rhonda Robb, Christine's assistant, who so promptly and graciously arranged my discussions with Chris and helped me get access to materials and team members at TeleChoice.

My special thanks and gratitude to Rashmi Doshi, CTO and co-founder of Everest Broadband, a building-centric service provider, who offered his valuable time for discussions on several topics and graciously answered the numerous questions I asked him. I would also like to express my great appreciation to others who have contributed ideas, written material, research material, comments, and suggestions in the course of writing this book—namely Katie Wacker of McLeodUSA, Jim Marsh of the The Management Network Group, Paul Roemer of Spectralliance, Bob Larribeau and Keith Mayberry of telecom researcher RHK, Norm Bogen of the Cahners In-Stat

Group, Tina Murphy of Forrester Research, the research firm Gartner Group, Eric Keith and Joel Ranck of Current Analysis, Rolf Brauchler, Richard DeSoto, Akshay Sharma, and Weicheng Liu. In particular, I would like to thank the various telecommunications analyst firms I worked with (named above), who were gracious and kind enough to allow me access to proprietary research material that they would normally charge several thousand dollars for. There are several others who provided insight and research material but wanted to remain anonymous and so I simply acknowledge them for their contributions.

Last but not least, I want to thank my mother for her loving support, for taking care of my daughter and providing me the necessary flexibility with my hours so that I could finish this book.

PART ONE

THE YEAR OF
THE CRASH:
A POSTMORTEM
OF THE
TELECOM CRISIS

WE BUILT IT AND NOBODY CAME

The setting for this headline story is scattered throughout the United States. Or more precisely, it's a worldwide phenomenon of which the U.S. bore the brunt. At one time we had hundreds of them, but now they number in the dozens. I'm talking about a group that narrowly escaped becoming an endangered species: the Competitive Local Exchange Carrier or CLEC. Dig a little deeper and it appears that not just the CLECs, but established giants like AT&T and Sprint have been seriously wounded as well, albeit not fatally. At any rate, it can be argued that no operator will emerge entirely unscathed from this meltdown.

Finding a starting point from which to survey and understand telecom's biggest stumble isn't easy. There are numerous examples of how the meltdown in the industry is rippling through the economy. A good deal has already been written about why so many companies went belly-up in such a short time. If you're in the industry, however, you probably don't think that the pundits have explained things in their true complexity. You probably believe that there are several as yet unexplained reasons for the meltdown, some well known and some not so well known. These are explored below, starting with the obvious.

I want to specifically credit TeleChoice for some of the key concepts presented in this chapter.

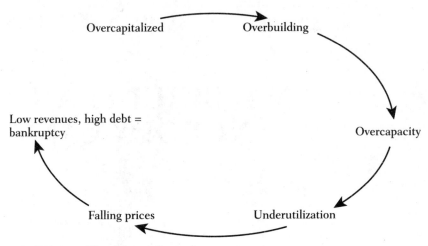

FIGURE 1.1   The pursuit of growth at any cost.

TABLE 1.1   Bank Loans and Debt Incurred by Telecoms. *Source: Thomson Financial (for 1998–2000).*

1998	$169 Billion
1999	$169 Billion
2000	$210 Billion
2001	$700 + billion (cumulative)

CREATION OF AN UNNATURAL MARKET ENVIRONMENT

Just six years ago, when the industry was jubilant over the newly enacted 1996 Telecommunications Act and hundreds of upstart telcos rushed to exploit the fresh investment terrain it implied, no one would have guessed what would happen by the turn of the century. Opportunities looked limitless and perhaps only the ILECs understood, to some extent, the nature of things to come. At the very least, nobody really understood the implications of the 1996 Act especially in terms of competition and investment. Somewhat uncharacteristically, we concentrated instead on what it was supposed to mean: lots of new choices, new suppliers, new applications. CLECs felt invincible behind the new protections afforded by the 1996 Act. All these dynamics fostered an era of exuberance on the part of telcos, investors, and shareholders, creating a world flush with capital—and replete with too many new competitors for it. Here was a classic unsustainable market environment, in which natural pruning had to occur before we could start to understand just what the new telecom market actually could be.

CLECS WERE CAUGHT UNPREPARED

A big mistake on the part of the CLECs was their failure to comprehend that they were newcomers on the scene and needed to invest time to stake out positions and build defenses for what was surely going to be a brutal battle with the entrenched ILECs. But time was short and budgets were high, so every CLEC thought there were enough customers to go around. Their attitude, roughly speaking, was the more competition the merrier (after all, who needed profits?). Few gave serious thought to identifying a market sector and positioning to win it against the competition—other than by lowering prices at the cost of profitability. CLECs did not foresee the problems they would encounter when they went up against the ILECs.

The CLECs were also caught unprepared to run a profitable business. Perhaps because of the seeming urgency to enter the market, or because they got distracted by the formidable response from the ILECs, or simply because they were too green on the playing field, the CLECs did not focus on developing clear business models for marketing, selling, and delivering broadband services. These companies spent on network build-out, without constraint and without plan, afraid most of all that they would be left behind and eager to get off the bench. Questions they couldn't answer included: what is the cost of implementing broadband? which services are ready to roll out? how should services be priced and billed? In short, they neglected to find out how they would make money as providers.

Rash spending without preparation or planning resulted in a huge hangover, from which the telcos are still suffering. Customers failed to materialize, funding dried up, and companies were left holding large amounts of debt that dragged them down.

THE TECHNOLOGY ADOPTION GAP

Another key point missed by the CLECs is that there's always a gap between the introduction of a new technology and user adoption of that technology. Demand for broadband services failed to materialize in the orders of magnitude that the CLECs anticipated. Small telcos in particular held out extremely high expectations rather than modest ones, and they spent heavily on network build-out projects rather than conserving resources for customer acquisition down the road. Indeed, what first seemed like insatiable demand begins to look more like a blip, which showed up on our radar screens as early adopters rushed to try new products. This early demand was not indicative of a continuous flow of mainstream demand.

The level of demand that the CLECs require is generated, not by early adopters, but by the mainstream majority. A transition from early adopters to the mainstream majority (where the profits are) occurs when the technology price stabilizes and application support arrives. At that point, the product or service is at a stage where it is safe to adopt, can be reasonably priced, and has clearly understood benefits and uses. In the case of broadband, the large majority of consumers never understood quite what to do with it. The supporting applications to spur demand just weren't there. While CLECs were building networks to carry large amounts of high-speed Internet traffic, small business users and household consumers were perfectly happy with their existing T1 lines and dial-up traffic. Today, in fact, most small businesses in far-flung areas (with as many as 50 employees!) are still using dial-up to access the Internet. What the CLECs did not realize, and what the ILECs unwittingly benefited from, is that users hardly ever get dazzled by "wow" technology to the extent that manufacturers and service providers do. That's the reason why venture capitalists are constantly searching for that elusive, hard-to-find "killer" application and never finding it—for the preponderance of new technologies, the killer app doesn't exist. When they do, more often than not, killer applications don't involve a radically new way of doing things, but instead involve a radically new technology that allows existing applications to perform better, cheaper, and faster. Consider, by way of a recent acute example, why 3G networking has failed to excite consumers. 3G, while admittedly a radically new technology, does nothing to enhance the applications we know and depend on. Instead, it introduces a new activity into the consumer's life (such as surfing the Web on a cell phone screen or downloading personalized information bits). If the consumer hasn't imagined herself engaging in that activity, she certainly doesn't feel the lack of it. This is much like saying that what we don't yet do is what we're not ready to do.

Several big European players (Deutsche Telekom, France Telecom, and others) are notorious for having already spent more than $100 billion for 3G licenses to provide wireless data services. These same players are now consigned to a considerable wait before they can reap profits from the investment. The "3G rollout" isn't happening on schedule. Moody's Investors Service recently reiterated its negative outlook on the European telecom industry, saying that cash flow from 3G "is very uncertain, both in terms of amounts and timing." The 3G dilemma—carriers are excited about it, users are not—can be explained as an adoption gap: the technology is there, but the users still don't know what they're going to do with it. Until 3G makes what we do every day "cheaper, better, faster," we won't require its services.

One exception to this pattern is the Internet, which was indeed a killer application in that it introduced a genuinely new way of doing things. But note that the Internet was around for a long, long time before it eventually became mainstream and finally exploded into our lives.

THE UPGRADE IN SERVICES WASN'T ACCOMPANIED BY AN UPGRADE IN THE NETWORK

To a large extent, the underlying infrastructure that enables broadband services remains old and costly. This manifested itself in the slow and expensive rollout of broadband services and created another significant barrier to user adoption. Today, broadband is burdened with a high price tag. Let's put it this way: the average user is reluctant to pay $49.95 for a DSL connection when the incentive for moving from dial-up is marginal anyway. And the average small business will not pay thousands of dollars for a fiber connection when a fractional T1 (at a fraction of the cost) is meeting its needs. To accelerate adoption in the absence of high demand, lower prices are an essential way to encourage users to switch. Unfortunately, in the case of broadband services, lower prices can only be realized if the provider is able to upgrade the network to reduce the costs of delivery. This didn't happen. As it turns out, the average variable cost of deploying a DSL line can be as much as $35 to $40 per month, not leaving much room to lower prices.

When CLECs understood that prices needed to be lowered to attract customers, they paid a high price for it: low profit margins followed by bankruptcy in some cases. Some, like Urban Media (now defunct), even offered *free* high-speed Internet access to amass a customer base. Not surprisingly, the company was not able to sustain this service and went under. With hindsight, what happened to the telecom industry seems inevitable. Later on in this chapter, we will show that this seemingly irrational behavior may in fact be a predictable stage in the life cycle of an industry. But first, let's revisit the economic damage caused by the meltdown.

Both the United States and Europe suffered from the intense price competition and lack of consumer demand for next-generation technologies, such as 3G. United States long distance companies, in particular, have suffered a great deal. European phone companies, such as British Telecom, which borrowed heavily to buy wireless licenses, are in deep financial trouble as well. Meanwhile, several dozen American telcos have filed for bankruptcy this year, and dozens more probably will by the time this book is published. Overall, the industry's debt looks not unlike a time bomb ticking. Telecom players in the United States and Europe have run up nearly $700 billion in debt, and analysts estimate that more than $100 billion in junk bonds will end up in default or restructured. Ultimately, say industry analysts, the telecom meltdown could be as costly as the $150 billion taxpayer bailout of the savings and loan industry in the late 1980s.

Because the telecom industry plays such a big role in economic growth, it is no surprise that the fallout has been felt beyond the borders of the industry, and may have potentially crippled the United States and European economies. Massive spending cuts by the struggling telcos have affected both the severity and length of the economic downturn, and may be one of the reasons for the current downturn. Major telcos have collectively laid off hundreds of thousands of employees across the country, exacerbating the affect of their rising debt on the stability of financial markets. This is how you get recessions, according to David Wyss, chief economist at Standard & Poor's.

VICTIMS OF THEIR OWN AMBITIONS

Teligent, Winstar, Northpoint, Rythms, Broadband Office, and OnSite Access are only some of the service providers that have failed. In the past year, these young CLECs saw their financing dry up and their businesses falter at an alarming rate. The aftermath can be felt in the equipment auctions taking place on the Internet and elsewhere. Literally brand new routers, switches, and computers have flooded the open market, fetching as little as 30 to 40 cents on the dollar. Several industry analysts and newspaper reporters have made it their business to poke around in the rubble and try to assess the damage wreaked on investors by the telecom bust that is shaping up to be one of the biggest financial fiascoes ever. "I don't know if there's a modern-day precedent for the billions of losses to investors from the telecom industry," notes Greg Dube, head of global high-yield investments at Alliance Capital, speaking to *The Wall Street Journal*.

THE BURSTING OF THE TELECOM BUBBLE

We all know about the bursting of the dotcom bubble. In fact, the telecom bubble was far bigger and, when it burst, it had much greater ramifications on the overall economy than the dotcom bubble, which was mostly contained within Silicon Valley. So, when did the telecom bubble begin to form, and why didn't we see it coming? The bubble actually had its origins five years ago in the 1996 Telecommunications Act, when Congress lifted restrictions on who could sell voice, video, and data services in the local phone markets. The Internet and wireless services were in their infancies and flush with promise. Hundreds of upstarts rushed to build state-of-the-art networks for the expected surge of demand, and incumbents responded by investing billions in their own wireless and Internet businesses.

Wall Street firms have made $7 billion in fees by raising debt and equity for these companies since 1995. But when the demand didn't materialize, and the Baby Bells proved to be tough, deep-pocket competitors, the start-ups found themselves with infrastructure they didn't need. Today, more than 95 percent of the fiber-optic capacity goes unused.

LACK OF ACQUISITION APPEAL

The shakeout is ongoing but, strangely enough, the predicted consolidation through acquisitions has not materialized. Northpoint Communications, PSINet, and Winstar are some of the better-known fatalities, particularly since they were also the high flyers of the new telecom era. None of these companies was acquired as a business, even though they had considerable worth in tangible assets and were fully operational. When Northpoint's assets were offered to bidders earlier this year, interested parties were scarce. One problem in this case is that Northpoint's business line was not unique; indeed the company provided essentially the same service as Covad Communications and Rhythms NetConnections, both of which have since declared bankruptcy and started selling off assets. Given Northport's complete lack of competitive differentiators, the final bid was a paltry $135 million dollar offer from AT&T, which barely covered the CO equipment the company had installed. Shareholders didn't get a penny, and investors lost their shirts on the stock.

This raises an interesting question: why are the struggling CLECs worth so little to acquirers, especially when they own such leading-edge technology and have—in some cases—laid out networks of some significant value? Why were more CLECs not acquired on the cheap, but instead forced to liquidate or declare bankruptcy? The answer is twofold.

First, a CLEC's worth lies in its operating model rather than its infrastructure. And efficiency is key to any operating model. Unless operational efficiencies are achieved, a CLEC can quickly become a losing proposition. This is exactly what happened with most CLECs who went under. The fact is that most CLECs never became fully formed businesses with viable operating models and business plans. PSINet, for example, never integrated many of the 74 smaller companies it purchased in recent years. These CLECs, although they were operating businesses, did not possess the business or financial savvy for the long term. ("Gold rush" is not the only metaphor for the late '90s: "shootout" is just as apt.) Acquiring such a CLEC spelled disaster, since the buyer would have had to spend more on the CLEC than he could ever get out of it. It quickly became apparent that if the acquirer wanted that business, it was easier to start from scratch than try to unravel the mess the struggling upstart was in.

The second part of the answer is that the cutting edge technology that the CLECs owned had a distinct lack of appeal. The large telcos already have an overabundance of it. It may not have been as obvious, but the big guys (the would-be acquirers) also engaged in unfounded build-out. The glut of capacity resulting from overspending during the last few years has made the telecom giants wary of investing in yet more infrastructure without first turning a profit. As a final insult, it quickly became apparent that anyone wanting to buy distressed assets had plenty of companies to choose from, and could buy on the cheap.

Experts agree that regardless of the next-generation equipment deployed or the visionary roadmaps, the assets of small telcos are pretty much worthless unless accompanied by a profitable business model or customer base that the acquirer can exploit for immediate returns. "These telecom companies are worth substantially more as ongoing businesses than in liquidation," says Aryeh Bourkoff, a telecom analyst at UBS Warburg LLC.

DEPLOYMENT STRATEGIES THAT WENT AWRY

If there is one thing the CLECs have learned, and learned well, it is that there is more to entering the market than pulling fiber, hooking up a switch, and furnishing a back office. The "build it and they will come" theory on which massive networks were built across the country—with heavily borrowed money—has instead become a cautionary tale. In the end, upstart CLECs simply could not generate revenues fast enough to cover the fixed costs of their upfront investments. Acquiring customers turned out to be more difficult and expensive than expected, particularly with the well-funded and crafty foot-dragging counterstrategies of the incumbent LECs. The ILECs were finding out that they could simply starve fledgling competitors to death.

Table 1.2 illustrates the amount of (net) investment and debt carried by some of the CLECs for every dollar of quarterly sales (based on 4th Quarter, 2000). As illustrated by the table, an anxious credit market was more threatening to some CLECs than others. Focal Communications and Choice One, for example, seem to have weathered the market pretty well for now.

TABLE 1.2   Investment, Debt, and Sales of CLECs

CLEC	INVESTMENT/SALES	DEBT/SALES
NorthPoint	$18.96	$20.31
Focal Communications	$6.70	$8.08
Choice One	$10.26	$15.15
PSINET	$6.40	$10.25

Although facilities-based carriers have been more threatened by the credit crunch than their nonfacilities-based counterparts, this does not necessarily argue against being facilities based. In Chapter 4, I'll show that being a facilities-based carrier in fact remains the most viable market entry strategy for carriers. The misstep most facilities-based CLECs made was in sinking huge amounts of capital into facilities deployment *prior* to signing up enough traffic to warrant the investment. When pressured to sign up customers fast, CLECs had nothing to leverage but venture capital.

A smarter strategy for facilities-based deployment, as I argue in the following chapters, is to lease facilities (equipment and networks) in the first year or two of operations, while focusing simultaneously on building a customer base. Base in place, you can productively begin thinking about ways to reduce costs and improve your bottom line. Fixed investment in facilities deployment is the natural next step. You can migrate from a lease-based, flexible financing arrangement to an outright purchase, thus creating for shareholders, in the plus column, the financial case for facilities deployment. Market entry strategies for young carriers are explored in greater detail in Chapter 4.

THE BLEC MARKET SHAKEOUT

Building-centric local exchange carriers (BLECs) did not escape the turmoil either. Early providers to focus on this market were young, private firms, and funding was key to their survival. In 2000, about $1.8 billion in capital was raised by these companies. Names that come to mind include Urban Media, Broadband Office, Everest Broadband, and OnSite Access. Larger service providers such as WinStar, Teligent, XO, and Allied Riser were also targeting this market segment. Today, Urban Media, Broadband Office, OnSite Access, WinStar, and Teligent are in bankruptcy proceedings, whereas Allied Riser seems to have retrenched and gone quiet, as though waiting for the storm to subside. (Note: Some consolidation did recently occur in the BLEC market. Cogent has bought out Allied Riser, e-Link has bought OnSite Access, and Yipes has bought Broadband office assets.)

Table 1.3 shows a list of providers in the building-centric space. Those in the left-hand column ceased to exist at some point after January 2001. The right hand column consists of companies that are still around.

TABLE 1.3   The BLEC Competitive Marketplace—Then and Now

JANUARY 2001	DECEMBER 2001
Allied Riser	Allied Riser (acquired by Cogent)
Advanced Radio Telecom	
Big Net	
Broadband Office	
Cmetric	
Cogent	Cogent
Comactiv	
Cypress Communications	Cypress Communications
Covad Communications	
Digital Broadband Communications	
DSL.net	DSL.net
e-Link	e-Link
Eureka Broadband	Eureka Broadband
Everest Broadband Networks	Everest Broadband Networks
Eziaz	
Genuity	Genuity
Harvard Net	
Internet Express	
Jato Communications	
Northpoint	
OnSite Access	Acquired by e-Link
Phoenix Networks	
PSINet	
Rhythms	
Telegent	
Telseon	Telseon
Urban Media	
Winstar	
Wired Business	Wired Business
XO	XO
Yipes	Yipes

One BLEC that has weathered the storm so far is Everest Broadband, though only time will tell if it can withstand a prolonged downturn. Industry insiders have praised the company, and its success story is apparent to the discerning eye. What is

Everest doing right? I interviewed company executives to find out. Everest's insights and best practices are captured in this book, along with those of other survivor providers.

Of course, everyone has a perspective on how and why the industry decline happened. Particularly in the case of BLECs, it is clear that service-oriented positions survived better than tranport-oriented ones. Rashmi Doshi's (CTO of Everest Broadband) perspective on the BLEC shakeout is described below:

> In the BLEC market place, there were three different classes of players to begin with. Very early on, a class of players emerged, mostly property owners, who thought there was a potential opportunity to start capturing communications revenue in multi-tenant buildings by charging for electricity and bandwidth. They thought they just had to put in telephone cable, and proceeded to do just that. A whole bunch of companies started, focused around this "communications" theme, which was purely a build-out focus, and had no idea or concept of the services involved. This led to a bunch of equipment companies starting up, which we will call the second class of players. These companies, well funded by venture capitalists, also did not have a lot of service focus, but instead focused on the deployment of equipment. The third class of players came late into the business. They were different, in that they took a service-centric approach, which was a market niche. This latter class has emerged relatively unscathed.
>
> In the mid 1999 to mid 2000 period, there was enough money to go around, and no reason for Wall Street or investors to distinguish between any of these players. Companies were expanding and trying to buy buildings and get into the business. When the capital markets started to tighten, companies who started out purely with a view of building out and deploying bandwidth found they couldn't really justify the capital expenses they were making. The equipment-oriented players found it tough because they didn't have a service focus. The companies that are really surviving now are the ones that have a service focus, a complement of services, for business customers. The capital is drying up, the venture capitalists are having to find out how to manage and drive the services business, though they don't have much experience in this area.

THE INNOVATION CYCLE

Christine Heckart, president of TeleChoice, Inc., a strategic consulting firm in the telecom industry, offers a more sobering and strategic insight into the recent telecom shakeout. According to Heckart: "We're going through a bad time right now, but we don't think it is new or the first time this has happened in the industry, though I think the swings are certainly wild." Heckart describes the current meltdown as a "thinning" phase of an *innovation cycle*, a framework developed by TeleChoice to characterize the phases that the telecom industry experiences every four to six years. "We are nearing the end of the *feeding frenzy* phase, during which time the strong, cash-rich companies will starve out the weak and cash-poor. The whole industry will thin out and slow down. And the cycle begins all over again."

So what does this framework suggest about the industry and how could it have illuminated what just happened? The four phases of the innovation cycle theory are explained below. Note that in this framework, the circle is never complete, because you don't start at the same point each time. It just gets a little bit bigger, because the industry is perpetually growing and changing in dynamics.

THE SLOW MOTION PHASE.   A small number of very large and powerful players (both equipment vendors and service providers, because this is a symbiotic industry) exist during this phase. This phase is characterized by very slow innovation and there is a lot of inefficiency exhibited by players during this phase. Service tends to be poor and prices tend to be higher because there are a few large players controlling the market. Because of this oligopolistic environment, entrepreneurial people in these large companies and elsewhere see gaps that they can exploit—gaps in prices, cost structure, services, customer support, and other less-than-optimal qualities. Because whatever they see is less than optimal, these entrepreneurs seek to exploit the opportunities caused by these gaps. This leads to the next phase.

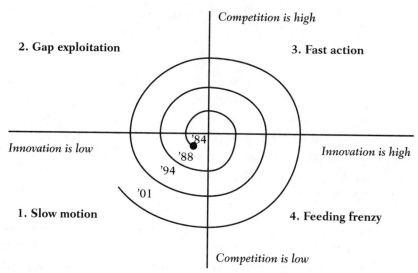

FIGURE 1.2   The innovation cycle. *Source: Telechoice 2001.*

THE GAP EXPLOITATION AND FAST ACTION PHASES.   Lots of little companies emerge to exploit the gap, both from the technology supplier and the service provider sides. Characterized by fast action, this is a chaotic and stressful time, especially for the incumbent players. A lot of new markets within the industry are created. You don't know which of these markets is going to survive and thrive for a long time, and which is a "flash in the pan" that will go away quickly. Many of the larger companies wait on the sidelines to see which, if any, of the gaps being exploited have the potential to threaten the core business or develop into substantial new markets. Once one or more new markets are poised to explode and pressure the incumbent markets, we enter the *fast action* phase. In this phase there is a high perceived risk of getting left out or being left behind. Service providers tend to buy their way into markets that look large and exciting, so that they don't get left behind—which is what happened to the ISPs in the past. This phase is characterized by a frenetic activity in gaining market share and mind share, although profits in each new area may be largely lacking. Companies that waited on the sidelines may now buy their way into the new markets, often paying top dollar. There are lots of examples of this in the ISP market, such as GTE buying BBN, and, more recently when Cisco paid a fortune for telecom gear companies.

THE FEEDING FRENZY PHASE.   Reality hits home during the feeding frenzy phase. The market can only support so many new ideas, products, services, and companies. Some make it, others don't. Those that sold out during the fast action phase probably got top dollar; those selling in this phase typically see increasingly smaller returns as the phase matures. The strong and cash rich starve out the weak and cash poor. The number of "threats" and "innovations" diminish substantially. The whole industry begins to thin out and slow down. Analysts believe that the *feeding frenzy phase* started around April 2000, that we're seeing the end of this phase, and are now headed into slow motion. The *slow motion phase* follows this, and the cycle starts all over again. Those who survive the *feeding frenzy phase* have a chance to play another round.

THE SLOW MOTION PHASE.   When you re-enter slow motion, industry dynamics have changed from the previous slow motion phase. It is more complex in some ways, and possibly simpler in others. All of the innovations of the past cycle have changed the dynamics of the industry, the way it serves its customers, and the way it operates internally. It is not a purely repetitive cycle. It is a circular pattern that, each time, is getting a little bit bigger, like a cyclone. Recognizing the pattern, and the phase, can help prepare companies to make the right decisions at the right time.

Who are the survivors in the innovation cycle theory? What do you do to make sure you are a survivor and can play in the next phase? Most likely companies that have built strong defenses to weather the cycle emerge as survivors. Several such strategies are discussed in Chapter 3. TeleChoice has developed a set of key characteristics that most successful companies share. Called the Seven Signs of Success, these characteristics are described at the end of this section.

Are these principles of survival applicable worldwide? While the innovation cycle theory is certainly applicable worldwide, the principles of survival tend to vary by country. It is hard to make generalizations on a worldwide basis, because competitive infrastructure drives the cycle. In a highly competitive environment, such as in the U.S. for instance, arbitrage play for a building-oriented BLEC may not be defensible. Consider, for example, the multiple connections or lines into each building. Someone else, most likely the incumbent, will surely have a lower cost basis than you, and will starve you out. In such a scenario, a BLEC would likely not survive using an arbitrage play. In an environment where competitiveness is much lower, however, you can make an arbitrage play and survive for a while until you can do something else. For instance, Everest Broadband has taken the approach of a service play, focusing on a vertical services market by targeting doctors' offices and other professional services. This has created deeper customer traction and stickiness for the company, although the creation of any new services model is risky and only time will tell if they can be ultimately successful.

It is sobering to think just how extreme the last round of the innovation cycle was. In the last round of the cycle, the stock market became a national pastime for the first time ever. The valuations were far and above what they should have been and had ever been. The *fast action* and *feeding frenzy phases* saw extreme and wild swings, both to the positive and negative sides. Executives began to manage stock prices instead of the company. Companies instituted stock splits to manipulate the stock price. Valuations, not business strategy, were the basis for most executive decisions.

THE FIVE FATAL FALLACIES OF NEW TELECOM VENTURES

Before delving into successful strategies, let's first identify the mistakes commonly made by young companies starting their new ventures. Sometimes dubbed the "five fatal fallacies," these mistakes set a company up for failure right from the beginning.

1. *The business strategy is the market or product strategy.* The results of this fallacy are the most deadly because the management team has no compass by which to guide the company and make good, timely decisions.

2. *A breakthrough improvement equals a disruptive technology.* Most new innovations improve the legacy, some even make new categories, but very few are actually disruptive. The strategies for each are completely different, and acting inappropriately causes severe problems. Only a handful of truly disruptive technologies have emerged in the past 100 years, such as fiber replacing copper, wireless versus wireline, packets versus circuits.

3. *Customer interest equals market success.* Customers are interested in lots of new stuff from lots of old and new vendors, but they buy from just a few.

4. *Incremental improvements (better, faster, cheaper) provide differentiation.* For a new venture, one or more order of magnitude improvements is essential, and these must be in areas very difficult for incumbents to duplicate without re-architecting or layering.

5. *Joining a "hot" category will result in higher valuations.* By the time a category is hot, you are too late!

THE SEVEN SIGNS OF SUCCESS

The most successful telecom companies share several key characteristics in operating style, business planning, and execution. If your company has all of these, you can turbocharge your performance and produce increases in shareholder value. If you have none of them, consider yourself imperiled in the telecom marketplace, and expect that your shareholders will eventually take their money and run. If you are like most companies, however, you fall somewhere between the extremes, and have already learned, in these turbulent times, that progress in these key areas is critical to creating shareholder value.

In trying to pinpoint the key characteristics of successful telecom companies, I spoke to several influential people in the industry to get their perspectives, and those conversations have shaped my formulation. What follows is particularly indebted to Heckart's analysis of what constitutes a healthy telecom company.

In working with young and old companies alike, TeleChoice has codified several outward signs of good internal workings. These are not guarantees, but since they can be "tested" by anyone inside the company or outside it, and can usually be ascertained in the course of a conversation with a company insider, they are highly useful indicators to employees, prospective employees, investors, and any one interested in the future success of the organization. These signs are outward symptoms of the deep internal workings related to a solid core business strategy.

1. *There is a **unique and defensible business strategy** that adheres to the vision and passion of the core team and the company's core competency.* Simply put, the company is a "strategy-driven organization," not a "strategy-challenged organization." The telecom industry is full of strategy-challenged firms: companies where decisions take too long and are often tentative or wrong. Their likelihood of survival is low. TeleChoice describes strategy-challenged firms as belonging to one of three types:

   · *Kitchen sinkers*—These companies suffer from a "goodness overload" and cannot make good and rapid decisions about growth because every direction seems equally appealing.

   · *Rat holers*—These companies mistake a product or market strategy for a business strategy. If market conditions change, they do not know what to do next. They often become Kitchen Sinkers until they can find the next rat hole.

   · *Backseat drivers*—They have a unique and defensible strategy, but it is continually debated and challenged by members of the executive team, thus hampering fast and effective decisions.

By contrast, strategy-driven firms understand that in today's environment mistakes can be deadly. Second chances don't exist or they are very costly. These companies understand the importance of staying focused, their decision making is fast and firm, they avoid the five fatal fallacies, and make a conscious effort to develop the seven signs of success. The underlying foundation is a deep and fundamental knowledge of what the firm is all about. It is the core area in which that company is the industry's best. It is the basis on which all decisions are made and measured.

2. There is a direct and well-understood relationship between the product(s) and the ultimate business goal.

3. The company offers a *defensible* order-of-magnitude improvement in at least one area important to the decision makers in the target market.

4. Company employees can describe what the company is all about in one simple sentence, and tell you why it is important in less than five minutes. (Most important, all the employees agree on this simple definition, and this is ultimately what aids rapid and effective decision making).

5. The company can show its differentiation and positioning in one compelling picture.

6. The offering(s) addresses a current, well-understood need in the target market (today's sales), while providing longer-term potential for dramatic change or opportunity (future sales and today's marketing).

7. Customers and analysts clearly understand the unique differentiation and can explain the value in a short and concise statement.

In contrast, in companies which do not exhibit many of these characteristics, if you ask ten executives what the company is about, you'll get ten similar but different answers that will take many minutes or longer to explain. The explanation will often be rooted in the company's market or customer base, and not in something intrinsic to the company, which it can justifiably claim to do better than any other company. The result is often a medley of offerings that take the company in many different directions and leave it vulnerable to better solutions from more focused competitors.

THE 1996 TELECOMMUNICATIONS ACT: POOR EXECUTION OR POOR GOALS?

Finally, we take a look at the role that the 1996 Telecommunications Act played in the telecom troubles. Congress and the FCC understood that expecting competition to emerge from CLECs building new, complete, finished telecom networks from scratch was unrealistic. As a result, the unbundling provisions of the 1996 Telecommunications Act were put in place. Seemingly, the underlying goal of the 1996 Telecommunications Act provisions was to have been the creation of an environment that would enlarge the addressable market by reducing the up-front fixed costs for new entrants to the market, making it possible for entrants to serve the market profitably and thereby fostering the kind of competition that would ultimately benefit the consumer and the industry. That's not what happened.

Unfortunately, the provisions of the 1996 Telecommunications Act fell far short of the promise, initially creating, but later stifling, competition. Under the provisions of the Act, in fact, there was little or no regulatory support for the CLECs to effectively combat the delay tactics of the ILECs. Keep in mind that while it sought to level the playing field, lasting competition, it seems, was never the purpose of the Act. The Act was designed to allow newcomers to put a foot in the door and investors to identify new opportunities. In short, instead of creating a long-lasting competitive environment, the Act inadvertently created a short-term boom, in which most players lost big, along with most investors. The demise of the CLECs was foreshadowed. Many opinions exist on the extent of the damage caused by the 1996 Act and the potential remedies. Heckart observes:

> If [the 1996 Telecom Act] was designed to create lasting competition, it would have split the ILECs into wholesale and retail units. But it was only designed to spur investment. Which it did, for a very short time, thus achieving the goal, although achieving little of lasting value.
>
> Had the Telecom Act put the ILECs into a wholesale group that controls the physical infrastructure, and made their survival dependent on selling lots of that physical infrastructure [unbundled network elements or UNEs] to retail-oriented companies, you probably would not have had the meltdown. But the ILECs retained control over the assets, and they very effectively strangled all of the CLEC new entrants into the market. It was not by accident, but by design.
>
> Other countries can learn from this. If there is one copper line to everybody's house right now, create a wholesale company out of it, and that company only survives if they sell to anybody who wants to buy. Should the wholesale side be regulated or not? Hard to say, but probably the competitive market they support will provide market incentives for self-regulation. Minimal regulation can be used for a transitional period to help offset severe and rapid negative impacts to the business units while they adjust. But in a competitive market, ultimately, even the monopolist can rely on supply and demand, not the government, to regulate its behavior.

Heckart's views on creating wholesale and retail business will probably never be validated, because the United States Congress has moved on to other more pressing issues in the wake of terrorist attacks and economic recession. However, the frustration caused to the competitive LECs by the "remonopolization" of the American phone system has not gone unnoticed by lawmakers. Today, more than a dozen states are considering sponsoring bills to bring competition to local phone markets. This legislation would require local phone monopolies (yes, the die-hard ILECs) to divide their companies into wholesale and retail operations.

Wholesale operations would be charged with running the local phone network that all phone companies must access to interconnect their customers. Its profits would depend on linking as many phone companies as possible to its wires. Retail operations would assume responsibility for the business of marketing and delivering local phone and other services, in competition with other companies. While not likely to result in the all-pervasive national restructuring envisioned in 1996, these attempts may provide incentive to the phone monopolies to open their access networks to smaller players and level the playing field.

Other industry insiders have expressed the view that the 1996 Telecommunications Act itself was not a failure. Rather, there was not enough pressure—financial or regulatory—on the Baby Bells to open up their markets and make the provisions of the Act a success. William Kennard, former chairman of the FCC, while speaking at a technology conference stated baldly: "The Telecommunications Act didn't create a free market." Arguing against wholesale deregulation in the telecom industry, he added: "That's a myth. What it [the Act] did was bring more balance to what was historically a monopoly market." The only way to foster real competition in local markets, Kennard believes, is through the intervention of government. In the meantime, Kennard further suggests, the ILECs are using the economic downturn in the telecom sector to argue that CLECs are not a viable business model and that the Bells are the only companies capable of building out high-speed residential networks. "I like to call that the 'seduction of monopoly,'" Kennard said, in rejecting the premise.

More than six years after the Act came into being, the tougher CLECs, badly battered, are gearing up for survival. They are trying to discover and hold on to viable business models amid almost-daily regulation changes, the near-inaccessibility of capital markets, bankruptcy filings by their peers, and the increasing power of the ILECs. Meanwhile, the FCC claims that the CLECs have made headway. Taking the view of a glass "half full," the FCC found in a recent study that CLECs saw a 93 percent market share increase over a one-year period, capturing 8.5 percent of the local market as of late 2000. The study found that one or more CLECs offer service to more than half of the nation's zip codes. Not surprisingly, the CLECs are heavy on business customers compared to ILECs: 60 percent of CLEC lines are strung to midsize and large businesses. This figure compares to 20 percent of ILECs' business, which is more heavily skewed to residential.

We should note that the study arrived in the middle of a current political debate over the level of competition in the local telecom market. Specifically, telecom players large and small are bickering over the Internet Freedom and Broadband Deployment Act of 2001, which favors the Baby Bells by lifting the prohibition on them from providing high-speed Net access, without requiring their open their markets to local competition, as specified in the Telecommunications Act. The Baby Bells, for their part, are taking aim at the cable monopoly—which may not be a bad thing after all.

FROM BUST PRACTICES TO BEST PRACTICES

No doubt the sorry state of the industry will have a profound impact on the landscape of telecom for years to come. And dozens of bankruptcies, in an industry once thought to be safe enough for the average naïve investor, are no longer unthinkable. Wall Street analysts and journalists have written sensational exposés pointing to the mountains of debt as the primary cause of failure, but the reality is that telecom companies failed for a host of reasons, ranging from poor management practices to insufficient cash to weak market positions.

PSINet, Northpoint, and Winstar are only three out of dozens of telcos that have gone under in the last year. Looking inside these troubled or failed companies gives us a sense of the types of poor judgment that some of the industry's executives showed. To that end, this chapter exposes some of the lapses in judgment that ultimately resulted in the demise of these companies. We follow with some forward-looking predictions by industry insiders. Building defensible positions to prepare for down markets is the subject of Chapter 3.

Some of the most common lapses in judgment that top telecom executives and their companies displayed include:

- Managing the company for short-term Wall Street expectations instead of focusing on the long-term prospects. AT&T's Michael Armstrong is a good example of a chief executive who did this.

- Going on wild acquisition sprees and amassing huge amounts of debt in the process. PSINet's CEO William Schrader masterminded several such acquisitions during the company's heyday.

- Investing billions of dollars up front in unproven infrastructure technologies that would take many years to pay back. Winstar and Teligent are both guilty of this.

- Filling idle capacity with unprofitable customers. Northpoint is a prime example of a company that lost sight of its customers in the mad rush to fill capacity.

MANAGING FOR WALL STREET, NOT SHAREHOLDERS: AT&T

Michael Armstrong initially started out with a good idea—build AT&T into a powerhouse that fills consumers' every broadband need—that might have had a decent chance of working, even taking into account the slowing economy, had Armstrong focused on the execution of the idea instead of on Wall Street expectations. No surprise to anyone, AT&T's CEO instead concentrated on building up the company to Wall Street analysts, creating ill-tested strategies around an idea that collapsed like a house of cards when the downturn hit. In fact, before the new idea was even validated, Armstrong started rapidly acquiring companies and restructuring divisions, paying out more than could ever be returned to shareholders and soon eating through $139 billion dollars of shareholder wealth. What's more, by scooping up these companies, AT&T lost focus and ended up with many different lines of businesses that had absolutely no synergy, thus ruling out the economies of scale achieved by resource sharing.

After this series of mistakes, it is unlikely now that AT&T, as it exists, will survive the current economic downturn. The company is struggling to get back on its feet, trying to realign its business lines, and also putting its assets on the auction block to gen-erate cash and help pay interest on the debt created from these acquisitions. In the third week of December 2001, the company announced the sale of its broadband cable operation to Comcast for $42 billion, in an effort to pay down debt and streamline its business lines to achieve some synergy.

GOING ON WILD ACQUISITION SPREES: PSINET

PSINet, Inc. was a company that ended its life $3.4 billion in debt, with annual interest expenses of more than $300 million, and a business that didn't generate any cash; a now-familiar story among the rank and file of the telcos. But not long ago, actually less than three years ago, the telecommunications upstart looked like a valuable player indeed.

One of the first to offer high-speed Internet service to cor-porate America, PSINet was a Wall Street hero, with market capitalization expected to surpass that of most *Fortune 500* companies. But PSINet's game plan didn't work out. Crippled by its $3 billion debt load, the company eventually ended up in federal bankruptcy proceedings.

Although it is certainly true that financial problems caused the company's decline, a closer look into the company's activi-ties shows that these financial troubles could have been avoid-ed and were the direct result of poor business judgment on the part of PSINet's executives. Upper management made several bad judgment calls during the company's heyday. The company went on a wild acquisition spree, even though big profits were years away, and amassed huge amounts of debt in the process. It built up Web-hosting centers in major cities including New York, Los Angeles, and London. Even as some members of the company's board of directors expressed concern about the debt, PSINet's CEO, William Schrader, "was more interested in building the infrastructure than in achieving profitability with what he already had," says Ian P. Sharp, a PSINet board member. With so much money available, Mr. Schrader wanted to get it before his competitors, according to Mr. Sharp.

Soon the company started seeing the first signs of trouble. This didn't faze the company executives, who continued to behave irresponsibly. Indeed, this was a time when the company could still have survived by making the right moves, such as getting acquired. But, according to reports, it appears that when rival telecom companies contacted the company to see if PSINet was willing to sell itself, Mr. Schrader demanded a price far above anything they were willing to pay, essentially spurning potential suitors. "When the stock was at $30 he'd say 'I'm a seller at $75.' When it went to $60, he said 'I'm a seller at $100,' " said a person close to PSINet. Mr. Schrader was sure his firm would be a survivor. At PSINet's holiday party in December, he told employees: "AT&T is going to go down, but PSINet will survive," according to a former PSINet executive who was there.

By September 2000, PSINet's debt load was hurting, sales were off, and the company was dealing with big costs just to keep its business running. In April 2001, when the company reported a stupendous $3.2 billion loss for its fourth quarter and said it had defaulted on several loans, Mr. Schrader was asked by the board to resign.

In the end, the company's debt discouraged potential acquirers, says an insider at PSINet. "You have companies like PSINet with a large amount of debt," he says. "Nobody in their right mind is going to step up and say 'I'll take that debt and buy the company.' "

Was debt the reason the company floundered? Undoubtedly. In the end, the debt was certainly a negative. But it's also obvious that the debt was only a symptom of the real disease, and that the lion's share of the blame must go to company executives for the poor business judgment they exercised, both in the acquisitions they undertook and, towards the end, their negotiating position with potential acquirers.

BLOWING MONEY ON UNPROFITABLE INFRASTRUCTURE: WINSTAR

It has been said that if there's one firm that epitomized the "I believe" mindset so prevalent on Wall Street in the late 1990s, it was Winstar, the former ski apparel shop that convinced the world that it could fly as a broadband company. On April 18, 2001, however, Winstar filed for bankruptcy, after accumulating more than $6 billion in debt.

Industry onlookers have labeled Winstar a victim of capricious capital markets. "Last year, the market was willing to invest $1 today for $2 tomorrow," says David Barden, who followed Winstar for J.P. Morgan. "Today the market won't lend you a quarter to buy a diamond mine."

But the problem goes far deeper than that. It is much too easy to blame the undoing of a company on capricious capital markets. Moreover, Winstar had a solid management team, with lots of industry expertise and billions of dollars in capital that the company raised over a period of five years.

Winstar made two critical mistakes, one political and one technology-centric. First, it played the dangerous game of making highly unusual deals with other service providers that had nothing to do with its core business of hooking up customers to its broadband network. Most certainly, these deals were made to make the company's earnings look good on paper. Here are some of the deals that Winstar reportedly made: In October 1998, Lucent agreed to lend up to $500 million at a time to fund the build-out of what Winstar grandly described as the "world's first global end-to-end broadband network." That so-called vendor finance arrangement allowed Lucent to bolster its own revenues by selling equipment to Winstar. Then, in December, Williams Communications said it would pay Winstar $400 million over the next four years to use its wireless broadband network. In return Winstar agreed to pay an even bigger amount back to Williams for the use of its fiber-optic cable—$644 million over seven years.

Supposedly, Winstar's core business was hooking up customers to its broadband network, not making one-time sales of capacity and equipment to other providers. But, like other CLECs, the company didn't break out its revenue. And since some of these deals were booked on the balance sheet, the net effect was that the true operating costs of the business were understated.

The second critical mistake the company made, which has gone virtually unnoticed, was in its choice of infrastructure to deliver broadband to its target market of office buildings. Winstar hitched its bandwagon to a technology so cutting edge and so expensive that it failed to fully realize the extent of the financial damage caused by the network build-out until it was too late. The company spent billions of dollars to build a high-frequency, fixed wireless network to compete with the ILECs— the most expensive yet of all last-mile fixed investments. In fact, this may have been what propelled executives to engage in somewhat shady financial arrangements to "hide" true operating costs from Wall Street. After investing the amount of capital that Winstar did, no company could hope to turn a profit for many years, even taking into account a high penetration rate (which didn't materialize) and marketing savvy (which Winstar executives had plenty of). Even bullish analyst Kastan didn't expect the company to make money until 2007.

Here's the reality: fixed wireless technology as it currently exists cannot be deployed profitably by a startup CLEC. The extraordinarily high fixed costs of deploying this technology from scratch make it extremely difficult to achieve profitability even at high penetration rates. So why did Winstar and other CLECs like Teligent invest in such expensive and questionable infrastructure? The most probable reason is that these companies were hoping that if they could capture enough business, a large incumbent or long distance carrier such AT&T or WorldCom just might pay a huge takeover price to rid itself of the competitive nuisance. Of course, they never expected to produce profits until some distant date.

In a footnote to the soap opera: on December 21, 2001, IDT Corporation announced that it had acquired almost all the operating assets of Winstar for $42.5 million. IDT will be using the company's network to augment its own long distance and fiber business. With a new management team and a new charter for Winstar, IDT may put the CLEC's fixed wireless assets to good use after all.

LOSING SIGHT OF YOUR CUSTOMERS: NORTHPOINT

Founded in 1997, Northpoint raised $1.2 billion selling stock and bonds to build high-speed DSL lines to sell to ISPs and small-business customers. In early 2000, the San Francisco-based company was valued at $6 billion and riding high. Northpoint executives fell into the same trap that the dot commers did: don't worry about profitability, build it out, get customers at any cost. Collectively, this was due to the stupidity of the venture capitalists, founders, and board members of these companies. When business customers failed to materialize, Northpoint started desperately filling its unused capacity with unprofitable residential customers at $29 per month. Here are some of the factors that may have contributed to this marketing stupidity:

· Customers had better availability.
· The technology fell short of what was promised.
· The $100 per month business market didn't materialize quickly enough to fill the capacity.
· Lack of support by ILECs led to deployment and provisioning problems.

If Northpoint had done the math up front, executives would have realized that when you add up fixed costs, such as depreciation on the DSLAM and CPE, on a per copper pair or per line/month basis, there are three elements of unavoidable cost:

· Monthly lease of copper pair from ILEC at $12 to $20 per line per month
· DSLAM depreciation charges (capex): $12 to $15 per month
· CPE (modem) depreciation charges: $8 per month

These three elements alone add up to $35 to $40 per month. Even if you ignore truck roll and provisioning and installation costs, you still lose money if all you are getting from the customer is $29 per month.

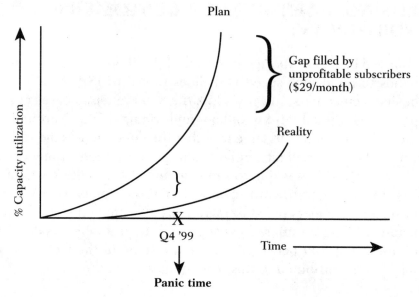

FIGURE 2.1   Northpoint: irrational move to fill the gap in business demand.

As the debt kept growing, the company was unable to keep up on its interest payments. When the only acquisition offer was withdrawn, Northpoint filed for bankruptcy.

WHAT NOW FOR TELECOM?

In spite of all the troubles that the telecom sector has faced, all is not lost. Industry analysts predict strong growth in telecom services over the next few years (see the last chapter for more projections from leading analysts). Although the telecom sector is likely to continue seeing upheaval and uncertainty for some time to come, end-user demand for telecommunications services remains robust. A recent study published, by the New Paradigm Resources Group (NRPG), a research firm covering the competitive communications industry, forecasts a $5 billion market for residential and business DSL within four years, *a nearly three-fold increase over current levels.* NPRG forecasts that revenues from business and residential DSL access will reach $5 billion annually by 2005, versus $1.8 billion today. That's pretty robust growth.

DSL, cable, and fiber are the primary contenders for broadband Internet communications to small business and consumers. Fixed wireless has pretty much gone by the wayside. DSL technology is the most pervasive, in that it transforms the existing copper phone lines connecting millions of homes and businesses to the global communications network into high-capacity data pipes capable of supporting broadband Internet applications. "Recent market turbulence and technical shortcomings have led many to misdiagnose the DSL industry as terminally ill," said Terry Barnich, president of New Paradigm Resources Group, in a recent online article that talks about the DSL study. "Last-mile broadband connectivity, however, can only come from a few competing technologies, and the market will support DSL's continued growth. The only question is, which existing business models can overcome technical hurdles, deploy DSL cost effectively, and meet the continued growth in demand for broadband?" In a nutshell, demand for DSL remains strong and yet unfilled.

Another report published by The Strategis Group says the CLEC industry is set to grow by over 350 percent. According to this study ("CLEC Strategies and Market Potential"), these events also paved the way for a brighter future, because capital will now flow more efficiently to a select group of strong, well-managed CLECs that have emerged from the pack.

"2000 really separated the wheat from the chaff," states Peter Jarich, director of broadband research with The Strategis Group. "Companies whose business plans essentially called for spending lots of money today in hopes of future profits were hammered by plunging market values and a tougher capital environment." Strong companies with cash on hand, solid revenues, and improving operating results, however—such as XO Communications, TimeWarner Telecom, and Allegiance Telecom—continued to attract capital and remain well positioned for the future, according to Jarich.

Indeed, these companies may have benefited from the troubles of their smaller brethren. Says Time Warner Telecom's CEO Larissa Herda: "There are positives for Time Warner Telecom as a result of the telecommunications shakeout." She says that they include a reduction in "irrational" pricing from others; a hindering of competitors from expanding into Time Warner Telecom markets because of their capital funding problems; a pickup of customers from competitors who are having financial problems; and a pickup of talent from competitors.

Driven by increased CLEC presence in local and long distance voice services, and continued strength in data services, CLEC industry revenues could grow by 375 percent over the next five years (see Table 2.1). In fact, CLECs have shown themselves to be more end-user friendly in the areas of provisioning, maintenance, and customer service operations, areas where their incumbent counterparts, the ILECs, have decidedly suffered.

TABLE 2.1   Total CLEC Revenues ($Millions)

2000	$12,865
2001	$19,333
2002	$27,696
2003	$38,769
2004	$49,738
2005	$58,781

Source: The Strategis Group, Inc.

Even the telecommunications industry's biggest names are going to have difficulty not getting singed in the years ahead. If industry growth continues to be modest, with all that capital being poured into new networks, one or more of the largest telecommunications players will end up in big trouble. There simply won't be enough profits to go around. The losers will see net income shrivel and their stock prices slide—and they'll have to cut expenses. "Not everybody is going to be a winner," says analyst Brian Adamik, of the market research firm Yankee Group.

SEPARATING THE WINNERS FROM THE LOSERS

So who *are* the winners going to be? This question is best answered by another question: Which sectors are the most vulnerable? Let us examine the sectors below.

Long Distance Carriers

The Interexchange Carriers that depend on the rapidly deteriorating long distance business stand to lose the most. The list starts with AT&T, but it also includes Sprint, WorldCom, and smaller players like Global Crossing and Level 3. The core businesses of the long distance carriers are deteriorating a lot faster than anyone expected. Long distance prices have continued to erode, tumbling more than 10 percent since 1997, and resulting in billions of dollars in lost consumer revenues for the long distance giants. And that price pressure isn't going to stop anytime soon. Lehman Brothers estimates that the long distance market is going to shrink more than 4 percent annually, from $78 billion last year to $63 billion in 2004.

Look for these companies to try to combine with others to gain economies of scale and shore up profits. AT&T, WorldCom, and Sprint are all candidates for acquisition or mergers. WorldCom and Sprint announced plans to merge last year, but the deal was blocked by regulators. Now, the two companies could end up being acquired by foreign telecom players or one of the Bells, after the local companies are allowed to offer long distance in more of their states. AT&T will almost certainly cease to exist as a separate company.

The Incumbent Local Phone Companies

The giant ILECs—SBC, Verizon, BellSouth, and US West/Qwest—look better positioned than the long distance carriers. Their local networks are costly and complicated to replicate, so today they're facing relatively little competition. At the same time, they're demonstrating that they're capable of taking market share away from the long distance companies. Verizon, for example, has swiped more than 1 million long distance customers in New York, and SBC has grabbed more than 500,000 customers in Texas in just two months. What's more, all three of them generate loads of cash that will help them finance expansion into new, fast-growth services. Verizon, for example, is expected to have a cash flow of $28 billion this year, more than either AT&T or WorldCom, according to analysts' estimates.

Another factor in the ILECs' favor is that the 1996 Act only applied to existing copper infrastructure. As regulations stand today, the ILECs are free to lay new infrastructure, such as fiber, without any pressures to share it with the CLECs. For the time being, unless or until new regulations are imposed, this definitely puts the ILECs in a strong position to move forward.

But the ILECs, not being a particularly aggressive or technically innovative group, are not completely immune either. They will soon face stronger competition from a newly emerging group of larger, more formidable regional CLECs that will be better equipped to introduce new services and support to a broadband-hungry community.

On the cable front as well the ILECs face a tough enemy—the deep-pocketed cable multisystems operators (MSOs). Facing a steep battle with the cable MSOs for the home broadband market, the ILECs will have to step up ADSL broadband rollouts or eventually lose the battle in their own backyard. The ILECs should start thinking about offering a bundled voice and data solution that matches the cable voice and data offering. The ILECs would also do well to start thinking of ways to get into the home entertainment market, such as video on demand; in addition to offering this over ADSL, the ILECs can take aim at the cable companies, which, so far, have a monopolistic lock on the cable broadband market.

THE COMPETITIVE LOCAL EXCHANGE CARRIERS

What is in the cards for the CLECs? Diminished access to capital and increased pressure to demonstrate strong operating results, combined with the necessity to achieve economies of scale and scope, *may* inevitably lead to consolidation in the CLEC industry. I say *may* because there hasn't been much evidence of consolidation so far, and it is entirely possible that, stuck with overcapacity and wary of the bad business models of smaller telcos, the larger CLECs or ILECs may well steer clear of going down the acquisition or merger path. There are likely to be more bankruptcies as well, and we are already seeing considerable scaling back in build-out spending. Almost certainly, a strong group of larger CLECs will emerge from the carnage; these companies will make it harder for the ILECs to thrive and prosper in the last mile. The most predictable winners are the regional telcos, which have zeroed in on niche plays, and well-funded national telcos like Time Warner Telecom and Allegiance, which look well positioned to weather the storm.

In the *CLEC Report 2001*, 14th Edition, NPRG points to emerging industry trends that suggest "shakeout" survivors will be prepared to succeed in the future. Steady industry-wide revenue growth, more realistic capital expenditures, and increasingly cost-effective technologies will speed the industry's recovery, while CLECs focused on regional markets will emerge in a strong position. In fact, going after niche regional markets is a winning strategy that more CLECs should adopt if they want to establish a stronghold for themselves. Several companies, such as BTI, NewSouth, Net2000, Network Telephone, and TelePacific have already seen the light and are choosing to remain regional players, focusing on their existing territory and customer base. These companies and others stand to build stronger customer relationships, increase brand awareness, and grow their businesses regionally.

"CLECs are obviously being transformed by the demands of the market, but the industry is hardly doomed," said Terry Barnich, President of NPRG. "CLECs will circle the wagons, regroup, and return to the fundamentals. They have a unique opportunity to learn from their earlier missteps and take corrective action. Customer service, revenue assurance, and market penetrations will be key for CLECs."

Increasing revenues will help speed the turnaround. According to the CLEC Report, CLEC revenues will top $93 billion by 2003, up more than 100 percent from $44.5 billion in 2000. Data services will continue to lead the growth, accounting for 58 percent of industry revenue in 2003, up from 46 percent in 2000. As revenues grow and capital expenditures decrease, the CLECs' bottom line performance will continue to improve. CLECs now operate roughly 16 million access lines in 1,500 cities, figuring most prominently in metropolitan areas, and targeting business customers with not just voice services or Internet access but enhanced services, such as Web hosting and VPNs.

Technical innovation will certainly play a key role in propelling the entire telecom sector forward. According to Bob Larribeau, senior analyst at RHK, technological innovation is one way to invigorate the sagging CLEC industry. New technologies will also reduce the future cost of doing business for CLECs, brightening their prospects for success. *Softswitches*, for example, were widely welcomed by the industry in 2000, and will increasingly be deployed to initiate new service offerings. Softswitches enable carriers to run voice and data services over the same network at one-quarter the cost of standard Class 5 switches. Had the CLECs been able to deploy softswitches during the booming expansions of the last few years, many would not be faced with the burdensome debts that have driven several to bankruptcy. Softswitch-based applications are discussed in Chapter 9.

CHANGING PRACTICES, REGULATORY AND OTHERWISE

Moving forward, the telecommunications competitive landscape will look a lot different from today's. The CLECs that survive, and any new market entrants, are going to find themselves operating in a much tougher capital market environment, with tighter access to capital, Wall Street's laser beam focused on earnings, and increased pressure to demonstrate strong top line and earnings results. Success on Wall Street will have to be earned the hard way, and no amount of pixie dust will work the charm. Rather it will be old-fashioned business models and sound business strategies that will do the trick, along with a little help from regulators.

In the midst of all these predictions and warnings, are there lessons for regulators as well? Yes, and for good reasons. Deregulation did not work. Not only did the competitive carriers suffer major setbacks at the hands of the incumbents, but the consumers did not benefit either. The majority of us are still waiting for broadband connections and reasonably priced innovative services to be delivered to us from a phone company that *we* choose. Today, fewer than 5 percent of U.S. households have broadband net connections, and the ILECs still control *over 90 percent* of all phone lines, with the competitive carriers servicing a mere 8.5 percent. State and federal regulators have the power to set the ground rules for a competitive marketplace. The question is: to what extent should they get involved, and how can they best help in fostering a healthy, competitive environment?

Business Week reporters Steve Rosenbush and Peter Elstrom recently wrote about eight controversial lessons that they claim will put the industry back on track and get it growing again. Most of these lessons are aimed at improving government intervention, and one of the primary benefits is to spur capital spending. For example, Rosenbush and Elstrom advocate that state regulators should slash subsidies and let the ILECs raise prices for basic service so that competitors will have an incentive to battle for customers. In rural areas, the government should subsidize the rollout of broadband Net connections to make it profitable for the ILECs, cable companies, and others to invest in more expansive networks. And in wireless, Washington should make more spectrum available to companies, such as Verizon Wireless Inc., that are ready to invest billions to deliver new services.

Given the right incentives, argue Elstrom and Rosenbush, telecom companies will begin to pour money back into the local residential market, broadband services, and wireless services. This would result in rebounding capital expenditures, which would put the ailing equipment manufacturing companies back on their feet. The following sections explore these radical ideas in greater detail.

· *Unless big changes are made, most consumers will not see the benefits of competition in local phone service:* [BW Recommendation: the current low price structure is irrational, beneficial only to the ILECs, and doesn't support competition. Why would the CLECs want to get into the residential market where there are no margins? Why not remove subsidies for most consumer households and let local phone companies raise their rates for basic phone service? This will encourage competitors to enter the market, consumers will get more choices and innovation, and prices will come down again.]

Raising rates for basic phone service is not likely to work for the following reasons:

While a profitable market should normally have the effect of luring more entrants into that market, this is not likely to be the case in the residential phone market. First, the real barrier to market entry for the CLECs has not been so much the low voice rates (CLECs never expected to make profits from the basic phone service anyway, but from offering a broad portfolio of voice and data services) but the foot-dragging and resistance put up by the ILECs. If rates are raised, this resistance from the ILECs will only escalate, because the ILECs will then have even more incentive to keep the competition out. Also, who suffers while we see if this experiment works? The consumer. Higher prices are definitely *not* in the interest of the consumer, who was to be the real beneficiary of the 1996 Act.

More important, the lack of demand by residential consumers for integrated services has held the CLECs back. The majority of the residential consumers are still happy using dial-up and POTS voice, and unless they start wanting (or needing) more, there *is* no market for the CLECs to enter. What is certain to increase competition is the much-anticipated arrival (proliferation?) of these converged voice and data services. This is what the CLECs excel at, and as residential consumers embrace converged offerings, CLECs will rush to compete for these customers. In this scenario, voice will cease to exist as a separate service and low voice rates will become a moot point. Conclusion: let the market forces of demand and supply do their work. No regulatory intervention is required.

· *The ILECs are more effective at stomping out local competition than anyone expected.* [BW Recommendation: State regulators should impose substantial fines on the ILECs, as much as $250,000 per offense, or they are not likely to change their anticompetitive behavior. Structural separation, which would split the Bells into separate retail and wholesale operation, could be an alternative remedy. The retail company would retain the Bell's customers. The other would sell network access on a fair basis to all rivals.]

Imposing steep fines may work to some extent, but it is not likely to substantially change the anticompetitive behavior of the ILECs. Moreover, these deep-pocketed carriers have the legal and financial resources to fight penalties all the way, resulting in prolonged litigation that could well take years. In the meantime, the ILECs will continue to frustrate and block the intent of the Act and effectively kill off their competitors.

The real remedy lies in the alternative: *structural separation*. Unless state regulators fundamentally alter the way the ILECs conduct business by implementing a structural separation like that now under consideration by some states, the residential market will, for the most part, continue to be locked up by the incumbent carriers. Structural separation is also likely to work because it provides the ILECs with a clear financial incentive to open their markets to competition.

· *The rollout of broadband Net connections is going to be slow, costly, and incomplete.* [BW Recommendation: Government should lead the broadband rollout, by creating a fund for network construction in low-profit margin areas, from the money saved by eliminating residential phone subsidies. Instead of giving all the money to the ILECs, let local phone, cable, and satellite companies compete for the subsidies.]

In a capitalist economy such as ours, the money follows the profits. Government intervention here will result in only half-hearted participation by the carriers, because of the low-profit margin potential in rural, undeveloped areas. Even if a carrier gets subsidized to build out the network, there are still ongoing operating costs to acquire and serve customers, and revenues are likely to be small due to demand for basic services only. The bread and butter of most CLECs is not basic service, such as high-speed Internet access, but a portfolio of advanced services.

The real incentive that government can provide is to pave the way for service innovation, starting with the densely populated, metro areas, where the profit margins are best. Innovative converged services will generate consumer demand and mobilize spending by consumers, generating profits and spurring capital spending and increasing competition.

Government can help by providing financial and tax incentives for carriers to upgrade older equipment and copper wiring to newer systems and technologies such as softswitches, building wiring upgrades, and fiber to the curb. This will encourage innovation and lower prices, as well as improve margins through a lower cost of delivery. Much of this innovation can be leveraged to the far-flung, remote areas as carriers look to expand their territories, thus making it easier for them to enter and serve low-profit margin areas.

· *The wireless industry is handicapped by the shortage of spectrum.* [BW Recommendation: Slaughter the sacred cows. Take spectrum away from the Defense Department, television broadcasters, and the satellite industry.]

After the September 11 attacks on the World Trade Center, the Defense Department is not likely to part with any of its wireless spectrum, which it will surely put to good use to battle terrorism. Taking spectrum away from the satellite industry, however, is merited and would solve some of the problem.

· *Given the chance, telecoms will litigate endlessly.* [BW Recommendation: Streamline decision making by regulators and courts, eliminating delays that have thwarted the Telecommunications Act.]

This is certainly a noble goal, but streamlined decision making and telecom regulators is almost an oxymoron. As long as carriers litigate, we are likely to see uncertainty and delays in decision making. The real trick is to take away the incentive to litigate, which is extremely hard to do in a capitalist environment.

· *The telecom regulators are reviewing mergers using old-fashioned criteria.* [BW Recommendation: Antitrust regulators should evaluate the proposed company's share of the total telecom market. Regulators shouldn't worry about whether a merged company would dominate one niche, but instead consider its share of the total telecom market.]

The U.S. telecom market is fragmented into sectors (wireline, wireless, local, long distance, and so on), with different carriers operating in each sector. It will continue to be this way until the sectors themselves merge, or all carriers start competing in all sectors. Until that happens, however, certain carriers have strongholds in certain sectors of the market, and it would be a mistake to ignore the sector-specific strengths of these players when reviewing a merger. Otherwise, carriers will merge to first gain strength in specific sectors and eventually to encroach on other sectors of the market, insidiously creating the very monopoly that regulators feared.

It is very likely however that the wire-line long distance and local phone sectors will start to merge in the not so distant future. I would recommend that reviewers consider not only the size of the sectors in which the companies currently operate and their ability to dominate that service sector, but also their ability to expand into other service sectors as a result of the merger.

· *Brand names and "one-stop shopping" are marketing myths.* [*BW* Recommendation: If companies want to charge premium prices, they must develop premium products.]

Elstrom and Rosenbush are no doubt talking about the ILECs and long distance companies, which have been slow to realize that old practices will not work in today's intensely competitive environment. These older, established companies (a good example is AT&T) have exploited their brands to charge premium prices, even though their products and service quality have steadily deteriorated.

The importance of branding cannot be minimized, however. Even innovative products tend to get commoditized over time. At that point, what differentiates one product from another is usually the intangible qualities associated with the company— the brand image, or reputation of a company to provide excellent support and deliver services reliably and on time. In an increasingly crowded market with remarkably similar products, these are the subtle advantages on which companies depend to acquire and retain customers. It should also be noted that a company invests significant resources into developing a brand image and merits some return on this investment through higher pricing. Regardless, customers will always be willing to pay more for quality brands and quality service. The key here is quality, of course. My recommendation would be that carriers need to focus on, and invest in, rebuilding these intangible qualities so that they do not risk losing customers.

One-stop shopping, or service bundling, is a phenomenon that customers have only recently been introduced to, and initial studies show that customers really like the convenience and lower price associated with one-stop shopping. A recent survey by IDC indicated that 58 percent of small businesses would like their services to be bundled. Understandably, bundling is being implemented to varying degrees of success, not so well by the ILECs (Qwest is an exception to this rule), and quite well by some CLECs such as Allegiance, McLeodUSA, and XO.

Carriers have embraced one-stop shopping because it reduces churn, but the delivery is less than perfect. What one-stop-shopping lacks today is the service associated with it. It is simply not enough to bundle all services on a single bill, and leave the poor customer to sift through the myriad items and prices on the bill. Carriers must recognize that once a customer has been signed up, the only way they can hang on to the customer is through both excellent customer service and continuous product innovation.

· *Open Internet standards really do encourage innovation and lower prices.* [BW Recommendation: Telecom-equipment makers and carriers must accelerate the deployment of new technologies based on Internet protocols and other open standards.]

The local networks are indeed based on proprietary technology. The last mile continues to be an obstacle to deploying multimedia services, such as video on demand and video conferencing, because most of it is based on old, proprietary technology. Newer technologies, such as softswitches, are not compatible with older, proprietary equipment in the network, so their adoption by carriers is delayed. The local network itself is old for the most part. It is impossible to obtain fast bandwidth at reasonable prices from the local phone companies. The main reason the last mile has not been "upgraded" like the rest of the Internet infrastructure is because it is controlled by the ILECs. Lacking competitive price pressures and not worrying about customer defection, the ILECs have had no incentive to upgrade to next-generation technologies. However, the last mile could indeed benefit from an equipment upgrade to a standard technology such as Ethernet or optical. Regulators must investigate potential ways to accelerate this change, whether through speeding up the approval processes for new legislatures promoting technological advances, or by using other measures, such as financial incentives to states that invest in an upgrade. This change will be welcomed by competitive carriers, equipment manufacturers, and consumers alike.

And there you have it. The rest of the book explores the best practices that telecom companies need to follow to be able to survive so that they can live long enough to play another round of the game.

BACK TO THE FUTURE: STRATEGIES FOR SURVIVAL

CHAPTER THREE

BUILDING A DEFENSIBLE BUSINESS

L ots of ways exist to build a business strategy and get the kind of defensibility that can help you weather a bad economic cycle. Half the telecom companies today don't have a defensible business strategy. They don't even have a business strategy, let alone a defensible one. There are several simple ways to create defenses by recognizing and then building on your deepest competencies. And there are some simple market truths that service providers can take advantage of. Transport or arbitrage positions, for instance, are not defensible, but service-oriented positions can be.

Companies that prosper in this new, capital-restrictive environment are those that will be able to meet traditional metrics and demonstrate a clear route to profitability. They will need to re-evaluate and reset their objectives and business goals to meet the requirements of the current economic climate. "The strongest and smartest players will focus on leveraging existing assets, building a solid customer base, and differentiating products with value-added services," says Dr. Judy Reed Smith, CEO and founder of Atlantic-ACM. To build defenses, CLECs in particular must carefully examine which situations to avoid and what strategies to emulate. Several CLECs are already doing this, and doing it well. CLECs like Allegiance Telecom, Time Warner Telecom, and others come to mind.

Singling out a few companies for special mention invites trouble. But when asked which overriding factor separates the winners from the losers, industry insiders focus on two things: having a defensible business plan that matches the company's budget and the discipline to stay committed to it. Sound management and financial discipline marks those companies that survive.

Whatever the markets, the CLECs that stand out have stuck to their business plans, calculated how much funding they needed for the markets they aimed to enter, raised that money, and focused on meeting their business goals to reach profitability. In developing their business plans, successful CLECs have also taken measured steps to build defensibility into their business models, using a number of strategies in their market and service selection.

To build a defensible business and market strategy that is hard for competitors to imitate, here are ten strategies you can use to strengthen your defenses against the competition:

1. Be different and be the best at something
2. Shore up your managerial talent
3. Know what your reach strategy is
4. Know how to scale your business
5. Institute smart financial strategies
6. Erect barriers to entry to choke competitors
7. Provide great customer service
8. Be diligent: Track market growth and churn
9. Create well-defined measures to assess and achieve business goals
10. Actively foster loyalty

BE DIFFERENT AND BE THE BEST AT SOMETHING—THEN STAY FOCUSED

This simplest of all strategies is the most difficult to attain. It is also vital to the long-term success of a company. You can build a defensible business position by being the best at something. This "something" is usually closely related to your deepest competencies. Most companies struggle because they don't understand that it is simply not possible to be the best at everything. Every new idea appears equally appealing and is eagerly investigated and often implemented, wasting valuable time and pulling the company further and further away from its original direction. Pulled in different directions, these companies are swayed from their core competencies, resources are challenged, decision making is slow and political, and the companies get stuck in a quagmire. The end result is usually dissension among the ranks, unhappiness among employees, lots of finger pointing, sliding revenues, and ultimately a new management brought in to institute a turn-around. By this time, it may already be too late to save the company, let alone catch up with the market.

To be successful, you must choose. You will have to pick something to be the best at, and then execute on it. Your entire organization must be focused on achieving excellence in one or two key areas. Your value proposition will also need to be centered on this key area, so that the market knows you for who you are. Above all, avoid having a value proposition that speaks of being all things to all people. For example, don't have a value proposition that speaks of scalability, reliability, intelligent switching, fast delivery, advanced services, higher profits, and on and on. Instead, focus on creating a defensible and achievable value proposition around one or two key areas in which you can shine.

This focus will result in your being the best at something versus aiming for more than you can achieve and diluting your message—which will surely result in an identity crisis for your company.

Questions that will help you focus may be "How do we grow?" or "How do we prioritize?" and "Why do customers buy from us?"

The key however, once you have decided what it is you want to be the best at, is in *staying focused*. You are going to succeed by being very focused and not by being all things to all people.

Companies that have lost their focus down the road have ended up with an identity crisis. To illustrate this point, let us examine the companies that have not been very good at staying focused.

AT&T, for example, is a company that has been particularly bad at staying focused. AT&T had the right idea in the beginning: "Create something new that changes the world in a way that other people have a hard time attacking—so that your defensibility goes up." There are two mistakes CEO Michael Armstrong made, however, in the execution of the idea:

- AT&T paid much more for the assets than could ever be returned to the shareholders. During the days of free-flow funding and excessive capital, when valuations were high and the opportunities seemed endless, most companies feared that they would be left behind. As a result, they ended up overpaying for assets. In the case of AT&T, part of their acquisition strategy was due to the "fast action" environment we were in, and part just to ill-considered decision-making.

- No time was given to allow the idea to mature. When the strategy was initially thought out, good execution around the "broadband solution for consumers" theme would have worked, but Armstrong gave the idea no time to mature. He let the organization actually create strategies around the idea before it had been completely formed and tested. In large part, this occurred because he was managing from the top and for Wall Street. Of course, Wall Street has financial smarts, but it is not a service provider. This eventually resulted in the meltdown of the company. Now AT&T is back to where it started, a "me too" company, but with a whole lot more debt and no defensibility in any of the business units. What exactly is it the best at today? AT&T has a reasonable brand and a large customer base, but it certainly doesn't have the best customer service or new assets around which to build. Going through an identity crisis, AT&T really has to start from scratch, and it is doubtful if the company has the time or resources to be able to recover from its mistakes.

Some very focused players (i.e., focused on being the best at something) are Williams, Time Warner Telecom, and Everest Broadband. The youngest of these, Everest, is best known for supplying and servicing the customer. Very early on, the company identified a unique niche in the multitenant space which it believed was not serviceable by volume players, such as the giant carriers. The company then proceeded with a laser-beam focus to build a value proposition that would appeal to this niche market: cheaper bundled services and better customer care. Everything Everest has done since has been around its central mission: to be the best at supplying and servicing this niche customer base. The company has not strayed from its tried and tested value proposition, and this has been a key factor in its staying focused.

Sometimes companies can start out highly focused but lose direction midway, as when transitioning or expanding from one market to another. The biggest danger at this point is losing focus. This often happens in boardrooms, very innocently. As one executive put it: "You call a board meeting and your investors say, 'Why don't you do this? Or do something else?' The CEO thinks, 'Here's my money talking.' And the message it sends to the CEO is: start thinking about moving the business in different directions."

The challenge facing many young telco CEOs is: How can we stay focused when we are being pulled in so many different directions, either by changing market conditions or by the conflicting interests of the board and executive team?

Here are some secrets to staying focused. First, *do not stray from your value proposition, whatever that is.* Unless you are diligent, straying often happens in subtle ways, usually through new products or services springing up (in your company) in response to perceived changes in the market or customer demands. This results in your marketing and sales messages changing; before you know it, customer perceptions have changed. For customers to stay loyal to you, it is important that you maintain the value proposition and company image that drew them to you in the first place. All new services or products that you offer to your customers must fundamentally enhance

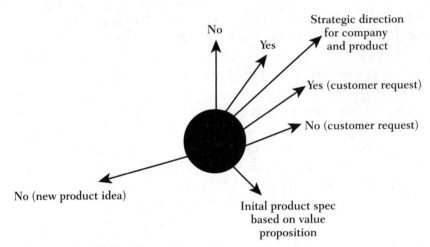

FIGURE 3.1   Know when to say "yes" to new product ideas and customer
requests and when to say "no."

your value proposition, not alter it. Creating an internal growth
vector can help you decide when to say yes and when to say no
to internal requests for new features or functionality.

Why is it so important not to stray from your value proposition? Your value proposition is the secret weapon that keeps
your customer base loyal to you even when the market forces
may temporarily shift against you. You are protected from small
adverse changes in the competitive climate as well. For
instance, small to moderate price cuts by a competitor, or the
advent of a slightly improved new service by someone else won't
cause your customer to leave you. That's why companies with
superior value propositions are able to hang on to their market
lead even when newer entrants arrive in the market touting
technologies that appear to threaten their leadership position.
The value proposition is also important because it ties directly
into superior returns for your shareholders. If the customer
believes your value proposition is superior to that of your competitors, the customer is more likely to keep doing business with
you, and you have extended the period over which you can continue to earn revenues from that customer. The key to creating
a winning value proposition is understanding your customers
better than anyone else—and using that understanding to refine
your value proposition and better allocate your own resources.

Second, *don't lose your entrepreneurial spirit*, which is another key factor behind your ability to stay keen and focused. Companies with entrepreneurial spirit are fast and nimble, always looking to offer a better deal to the customer. Such companies also have a strong survival instinct. Everest Broadband is an example of extreme entrepreneurship. While competitors in this industry have been tumbling, Everest, which began operating in January 2000, is a survivor. It has signed deals to bring its telecom services to 200 buildings, including such high-profile landmarks as the Chrysler Building in New York City and the Merchandise Mart in Chicago. It continues to roll out new services, competing with established giants such as Verizon and AT&T. After establishing its presence, Everest has stayed focused on building on its existing base of paying customers, expanding its base as well as getting its existing customers to spend more on Internet telephony, customized high-bandwidth service, and other leading-edge offerings. Everest is also trying to buy several competitors to beef up its offerings, but only at bargain prices and when the purchase enhances their core business.

SHORE UP YOUR MANAGERIAL TALENT

How much experience does your management team bring to the table? In the end, no matter how good your business and product strategy, how far you can go with the backing you receive boils down to simple good management. In other words, experience counts, particularly at the management level. "You can have the best business model in the world, but whether you have the management team to execute that business model, and really stick to it, is really a make-or-break situation," notes Atlantic-ACM's Regas. High-quality senior managers who have *been there, done that* before are invaluable assets to a company. These forward-looking people can predict and often pre-empt the negative results of shifting events in a rapidly evolving marketplace. For this reason a number of CLECs are starting to recruit seasoned executives who previously worked for an ILEC, IXC, or first-generation CLEC.

When it comes to local service, prior telecommunications experience is a huge advantage, considering the complexity of networks, with their legacy equipment, and the complicated relationship between ILECs and CLECs. You want to have executives on your team who have seen it all before and know what do.

Experts believe that experience is vital in shaping long-term strategic focus. "Smarter and more experienced management teams that have been in the business longer realize that money isn't going to be flowing in all the time, and it's got to be a long-term plan," New Paradigm Resources Group analyst Liz Singleton says. "A lot of these executives go from company to company. Some might say, 'You can do well one time and then things don't work out so well the second time,' but it's the people with good track records who lead these companies and do a good job."

KNOW YOUR REACH STRATEGY

A very important part of building a strong and defensible business is preparing early on to establish your presence or *reach* in the marketplace, and selecting the right channel or vehicle to carry your message. Reach provides a vehicle to carry your company's message, products, and services to the marketplace. You can establish reach or market presence in several ways, using branding, partnering, global presence, or other strategies to create a foothold in the market. For instance, you may decide to offer prepaid calling card services in developing countries in the Asia Pacific region. In this case, your vehicle to carry your message and products is your distributors or partners in these countries. They will, to a large extent, influence how successful you are in creating a foothold in this region. In another instance of using partnerships to create a foothold, you may decide to partner with a known national CLEC to distribute your videoconferencing services. There are many other ways you can create and demonstrate reach. Established service providers sometimes have an edge because they have the option to create low-cost reach vehicles through existing customer relationships. Examples include using an existing sales force to deliver new products and services to customers, or using an existing network to create new services. MCIWorldcom, for example, emphasizes its reach when it says that it owns the farthest-reaching IP network to carry its services. The network serves as its reach, on which it delivers new products and services. Verizon emphasizes its reach when it says that no other company can reach as many buildings and homes. You can establish market presence through a number of reach strategies. The most pervasive ones are described next.

COMMUNICATING AND SELLING THROUGH BRAND

Reaching your target market and potential customers through your brand can be an extremely effective reach strategy, although a prohibitively expensive one. Brand is often used by established carriers to reach their audiences with new products and services. Several telcos have used branding to increase customer awareness of existing and new programs, change company image, and introduce new features. Brand, for example, is an important part of Verizon's reach strategy. Using a highly visible national branding campaign to reach existing and new customers, the company introduced new wireless and single-rate services, which resulted in a very high degree of effectiveness.

In the case of XO Communications, the company used branding to inform the marketplace that it had expanded to become a national provider of integrated communications services. XO used its branding strategy to drive awareness of its new expanded flat-rate, bundled pricing for all services—local, long distance, broadband Internet access, Web site—combined with a single bill, 24/7 customer service, and a three-month guarantee.

In another example, McLeodUSA uses its Yellow Pages directories to reach customers and heighten awareness among them of the company and its services.

COMMUNICATING AND SELLING THROUGH PARTNERS

This is a rapid method for reaching your customers, and nearly all telecom companies, whether startup or established, practice it to some extent. Some service providers partner with their customers, while others co-partner with other providers in the value chain to deliver integrated services to end-customers. For example, Level 3 sells colocation gateways, softswitches, and core transport services to major carriers. For efficiency, these sales take the form of integrated relationships rather than isolated transactions. Global Crossing offers its global fiber network to partners for building services. Metromedia's metropolitan dark fiber business provides the infrastructure for colocation companies like AboveNet and others to build on. In this case, Metromedia owns its partners. For most service providers, it can make more sense to partner in a symbiotic manner with other players in the value chain to maximize reach and deliver integrated services profitably. Reaching customers through partnering is discussed in greater detail in Chapter 6, which investigates partnering alternatives.

You can establish market presence through a number of ways, so you will want to analyze what will work for you in your particular scenario. Which reach strategy is going to be most effective for you can depend on what type of service provider you are. For example, network service providers often choose network completion as a reach strategy. Transport providers often partner with their customers. Building-centric service providers utilize partnerships with real estate companies and landlords to allow them fast access to tenants. And regional, as well as established, carriers often depend on branding to introduce new services. In each of these instances, the service provider's business goals and management team capabilities usually influence its reach strategy.

KNOW HOW TO SCALE YOUR BUSINESS

For those service providers who have successfully implemented a niche market entry, the next point of expansion to larger or related niches is the most coveted, yet the most dangerous. This involves scaling your business to address a larger segment but not losing the individual touch you have built into your services.

COMMODITIZING FOR THE MASSES VERSUS MASS CUSTOMIZATION

Here's the dilemma most niche players face. If you're playing a volume game, then the next customer has to look a lot like the previous customer. The large carriers excel at this. In a typical niche, the customers are geographically concentrated, yet not very alike in preferences, and therefore not disposed to mass-produced products and services. These customers are cost sensitive, yet want individualized one-on-one attention. For this reason, large carriers usually stay away from niches. Niches require a level of individual customization that is not easily served by a volume player, making these areas the ideal market space for young telcos to bring their solution under the radar screen of the large volume players.

Now the niche player starts thinking about scaling up and growing because the niche, by definition, is limiting the scope of what it can do. To expand, the niche player must take its technology and move it to larger, related niches. *The danger is that the niche player tries to play the volume game by cookie-cutting a solution based on the least-common denominator.* This can kill the company. Many young CLECs, growing out of their entry niche points, have made this mistake of cutting costs and commoditizing a solution for the masses, when *mass customization* is really what is required. What is mass customization? Mass customization is a technique that enables you to differentiate yourself by optimizing your products and services for specific vertical applications. It brings the attention and detail found in custom-built products or services and couples it with the speed, quality, and economic benefits of mass production.

For example, if your niche consists of professional services businesses in multitenant buildings, then your bundle of telecom services should be tailored to fit the needs of each unique type of business, whether a doctor's office, or a law firm, or an accountant.

Leading companies, such as Dell and Fidelity to name a few, have successfully applied mass customization techniques to establish product or service differentiation in their respective markets. Claudia Bacco, vice president at TeleChoice, notes that successfully leveraging mass customization starts with recognizing the opportunity to apply service dimensions that will enhance the basic service offering. With telecommunications services, this includes understanding the various technical requirements of the applications, as well as the business needs of the user.

The hard part of implementing mass customization is having the financial and technical means to meet customer expectations across the board. Although corporate customers may be willing to pay a premium price for a first-rate telecommunications service, the more budget-conscious buyers are looking for a cut-rate deal—with service expectations to match. But no matter what level of services businesses pay for, all business customers want their money's worth. And that is the key—to make customers feel that they are getting their money's worth through products and services "touched" or customized for their individual needs.

ESTABLISHING UPSELL CREDIBILITY

Another way to scale the business is to get a strong foothold and sell existing customers more and more services. This can be either a perilous step or a way to establish credibility and gain customer loyalty and, consequently, additional revenues. You can establish a form of trust with your customers through asking them to pay only for upgrades that have meaningful value to them. Some companies make the mistake of selling upgrades that don't contain visible or functional improvements that customers find valuable. If this happens, the customer gets suspicious and wants to pay less for the upgrade. When this happens, you have lost upgrade credibility with this customer. This customer may even decide to switch to a competitor. Some experts believe a "gee whiz" factor is necessary to charge for an upgrade. In other words, the key to selling more services to existing customers is to introduce an element of dramatic improvement in service features or performance that will get the customer's attention.

America Online (AOL) is a company that has applied this principle very well, as it tries to up-sell dial-up customers to cable modem. AOL customers can readily see the increase in speed. They are happy because the connection is always on. It is indeed a boost in productivity, and so yes, they're willing to pay $39.95 a month (versus $19.99) for the upgrade.

INSTITUTE SMART FINANCIAL STRATEGIES

During good times, spending comes easy. During lean times, which we are experiencing now, companies put off spending even on necessary items. Both spending strategies are extreme and injurious to the long-term health of a company.

SMART SPENDING STRATEGY

The smart spending strategy requires that, regardless of the economic climate, a company should be continuously managing its budget and scanning the different parts of its business for both financial drains and rapid revenue spurts, with an eye to diverting money from unprofitable areas into profitable ones.

During periods of extreme economic contraction, when quick decision making is further compressed and rapid judgment calls are required, it can pay to adopt a "feed what works, starve what doesn't" tactic to keep revenues flowing and expenses down. This tactic requires adopting a short-term view of your return on investment. Quarter by quarter, you will need to be on the lookout for what works, because you will need to feed these areas. One way to achieve this is by drawing a line in the sand. Set revenue targets for each unit by quarter. If, at the end of the quarter, you find that you have exceeded the sales quota for a particular unit, you will want to feed it. If, however, you're way below it, you're going to have to make some hard decisions about how much time you will give it before shutting it down.

To find examples of this in the telco industry, one only has to look at the rapid twists and turns in the market entry strategies of several CLECs. One executive talks about his company's decision to pull out of residential buildings: "We were successful in pulling in this portfolio of residential buildings. We had a couple of buildings where we started to go and build high-speed data infrastructure. But it took resources—time, energy and money—we decided to take residential off the table." This is the nature of the fast decisions that are required to survive in today's unforgiving environment.

The "feed what works, starve what doesn't" edict is tactical in nature, designed to keep revenues flowing, and can work well in extremely tough capital environments. Longer term, this strategy is risky, because it can distract the company from its strategic direction and often prematurely kill-off promising projects that require nurturing over two or more quarters to produce profits. Indeed, many business units require longer than a single quarter to start producing a profit; you would be ill advised to starve these business units in favor of others that start turning a profit sooner. What is required, however, is a keen ability to distinguish between bad bets and good ones in a particular economic environment. Too many companies send good money after bad. Utilizing a smart spending strategy will help you identify and stop financing "bad bets" before you lose your shirt. If a particular business unit is not thriving, it is a better bet to let it go and come back to it later when the economic environment or other conditions change.

TAKE CARE OF YOUR BALANCE SHEET: SPEND WISELY

Cash rich companies have some of the best chances of survival. However, if a company does not do other things to build defenses, then conserving cash is not even applicable.

Over the last several years, the main focus has been the income statement. For instance, if you looked at the income statement of many CLECs, revenues were growing and shareholders felt good. No one really thought of the balance sheet, because capital was free. But when capital markets tightened and investors started to examine the balance sheets of these CLECs, they saw that, although these companies were operating, they were, for all practical purposes, insolvent: they had $500,000 in the bank and $20 million in current liabilities. In today's tight capital environment, the balance sheet matters more than ever—because running out of cash means going out of business, no matter how good your income statement looks.

So, your income statement may look great, but because your debt-to-equity ratio is what can hurt your business, it is your balance sheet that you need to worry about. The biggest concern among withering CLECs is their balance sheets. High cash-burn rates were a necessary evil during the boom years, as up-and-coming telcos tried to grow at all costs. But when capital-markets funding dried up, most CLECs found that they had burned through all their cash or were in danger of doing so—yet they still faced staggeringly high interest payments on their debt, which only compounded the bad situation. CLECs are learning, some too late, that you can build a defensible position by simply conserving cash to help you get through the economic downturn that inevitably follows the boom years.

In fact, survivor CLECs are those that have tended to their balance sheets. Put simply, these companies have used their cash wisely and kept their debt-to-capital ratio modestly low. Dallas-based Allegiance's smart spending may ultimately make it succeed. While rivals were expanding rapidly with the expectation that the capital markets could be tapped at will, Allegiance took a more sensible route. If you examine the company's balance sheet, you will find that it sports a good-looking (by that I mean conservative) balance sheet. The company's debt-to-total-capital ratio of 38 percent is very low compared to the 50 percent or higher ratios of its rivals. What does this mean for Allegiance? The company has adequate cash on hand to withstand the slowing economy and will most likely emerge a winner when others have fallen by the wayside.

Another plus is that having a strong cash position and a relatively low debt-to-capital ratio can help you withstand the pressures of short-term operating losses, something that every startup CLEC has to contend with in their early years.

ERECT BARRIERS TO ENTRY TO CHOKE COMPETITORS

Barriers to entry are highly desirable but can be difficult to erect in the telecommunications market space. Erecting barriers is desirable because early entrants usually have a significant market advantage over late entrants and the potential to make it difficult for later comers to carve out a significant market position. Barriers to entry have the effect of delaying the inevitable rise of competition in a profitable market by making it more expensive for new entrants to enter the market. This causes their temporary retrenchment, which gives the early entrant valuable time to grow its customer base and strengthen defenses.

Retrenchment usually occurs because the higher the investment to enter a market, the higher the risk of failure. Usually, barriers to entry become stronger the longer the early entrant can keep competitors out, amass a customer base, and create trust with customers. The residential telephony market is a great example of a market space in which the incumbent phone companies, cable companies, and AOL, the incumbent Internet service provider, have created effective barriers to entry: the first two through a formidable last mile build-out of copper and cable infrastructure, and the latter through an aggregation of 30 million customers and a deep stickiness resulting from trust and value. Assuming you are an early entrant in your chosen market, there are several ways you can erect barriers to entry:

· *Your target market itself can be a barrier to entry.* For example, the nature of niche or small regional markets is such that there is limited potential for profitability when customers are shared among more than one or two service providers. This characteristic itself serves as a disincentive for new entrants when a service provider is already serving that market. Moreover, niche markets require individualized one-on-one marketing that can quickly foster relationships between you and your customers, creating the kind of deep traction that cannot be easily overcome by a new entrant.

· *The nature of your company's reach can also erect barriers to entry.* In the case of the ILECs, their last-mile build-out, which serves as their reach for creating products and services, has erected significant barriers to entry for upstart CLECs wanting to serve residential consumers. Companies such as Everest have created barriers to entry through the tight partnerships they have formed with REITs and building landlords. A company's brand, if established enough, can also serve as a barrier to entry.

- *The aggregation of large numbers of customers can be a barrier to entry.* America Online is an example of a company that has erected huge barriers to entry for upstart Internet service providers. The growth and profitability advantage of the 30 million AOL subscribers has all but squeezed out marginal players in the battle over residential supremacy.

- *Trust can erect barriers to entry as well.* Trust locks in customers. No matter how compelling the new entrant's service offering, it will prove difficult to switch customers from a trusted relationship. Moreover, customer acquisition now becomes much more expensive, reducing profitability and driving away all but the most determined and deep-pocketed entrants. Trust is built by establishing credibility and closeness with customers, primarily through perceived value on the customer's part and customer relationship management on your part.

PROVIDE GREAT CUSTOMER SERVICE

Most service providers, when asked what a carrier's number one priority ought to be, point to the ability to provide total sales and support to the customer. But most service providers still do not understand how vital customer support is to building a strong and defensible franchise. At the end of the day, networks tend to be very similar. From an expense point of view, you have spent a lot of money, but you really have to have a good customer support and help desk that can solve the customer's communications problems, and a customer service organization that understands how to manage the product being sold to customers. These are the areas truly worthy of investment, because they will pay you back in good measure. A common mistake CLECs make is to get hung up on technology. Technology is used to reduce the cost of offerings—but if you don't have really good customer support and management and the automation to handle it, the rest is pretty useless.

When asked, many survivor CLECs attribute a good part of their success to their early attention to customer support requirements. Today, even with spending coming under pressure, Everest remains steadily focused on the customer service and operations support side of the business. The company started out by trying to make sure that, as much as possible, things were automated and managed early on. Says Rashmi Doshi, the CTO of Everest: "We didn't necessarily go out and invest in huge support systems up front, but recognized early on that we had to put it in as we grew the network. By factoring this into our initial planning, when the systems had to be put in place, it wasn't an afterthought but a well-planned move that had already been blessed."

McLeodUSA is another example of a CLEC that built its reputation around customer service and provisioning. Any company can get the latest technology, according to Bryce Nemitz, VP of Communications at McLeodUSA. What sets one apart from another is how effectively it operates. "It comes down to how customer-efficient your network is. Do you have the quality people and quality system to provide great customer service and provisioning?" he asks. "If you do that well, the efficiencies of the network pay you back."

Two ways that you can enhance the customer support experience are described below:

- *Web-enable your customer care.* Much has been said about expanding your business to the Web. The sheer volume of residential phone subscribers has probably made it impossible, at least for now, for local phone companies to e-enable their business for any but the most rudimentary tasks. For many other types of businesses, however, moving customer care to the Internet has resulted in significant savings and operational efficiencies. Cisco has perfected the art of Web-enabling its customer support division. In the process, Cisco has saved billions of dollars and increased customer satisfaction by several orders of magnitude.

  Web-enabling customer care makes sense for carriers serving small- and medium-sized businesses. In particular, managed service providers can augment the efficiency of their customer service operation by moving some of the management, reporting, ordering, and billing components of their customer care to a password-protected extranet that offers secure access to customers. Customers can use this Web interface to request additional services, change the parameters of existing services (such as requesting additional e-mail accounts or remote access VPN for new or temporary users), view SLA reports online, view bills online, make payments, and so on.

· *Be proactive in your customer care.* Proactive customer care means analyzing and improving the different components of your customer service operation so that they come together to create an overall experience that enhances customer care. This involves periodically analyzing existing customer provisioning and service activation processes, anticipating future service problems, and correcting them in a timely manner. For example, on a periodic basis, you will want to select a sample service territory within your target market and analyze the types of problems these customers experienced with their broadband service over the past 90 days. Use this information to predict and pre-empt potential service and staff problems that may not be obvious but that result in lost customers. If customers are complaining about losing connections often, this could point to a network-switching problem, be related to the ILEC's last-mile facilities, or be coming from another source.

Analyzing a group of service territories can provide valuable insight into whether certain service territories are underperforming when it comes to resolving customer complaints, and can point to deficiencies in service staff capabilities or territory-related network problems. You can zero in on the heart of the issue by identifying and tracking every step in the problem reporting and resolution process, until you isolate the problem and fix it.

BE DILIGENT IN TRACKING MARKET GROWTH AND CHURN

Companies that have a steady and growing customer base have the best chance of survival. There are two components to growing your customer base: growing your addressable market share and keeping customers from leaving you. Growing your market share (gaining new customers within your addressable market) helps you increase market size and revenues, and customer satisfaction keeps them from leaving you.

If you want to build a strong, steady business, you must institute programs to accelerate both market growth and customer satisfaction. Your ability to retain customers is your strongest weapon against the competition. Of course, the types of customers you want to attract and retain are the profitable ones (as illustrated in the Growth section below).

In the telecom industry, typical ways to measure growth rates are by tracking increases in account sign-ups, call volumes, new product revenues, prepaid calling, and so on. Customer satisfaction is typically measured by tracking churn rates. Below are some specifics for tracking growth and churn.

GROWTH

Many telcos do not understand how to measure growth in a meaningful manner and are frequently misled by the numbers generated by marketing campaigns. Growth is meaningful only when it contributes to higher revenues overall or translates into greater profitability for the company. The key in tracking growth rates is not to get sidetracked by absolute measures, but instead focus on the quality of the growth and the effect it has on your top and bottom lines. Is the increase in new accounts actually contributing to top line growth (i.e., higher revenues)? At what expense are you acquiring new customers? Are these customers really new and are they profitable? Are you cannibalizing one area of your business to create growth in another? Is your bottom line hurting even if your top line is growing?

I recently heard an interesting example of how growth can be misleading, and how the quality of the customers can affect the top and bottom lines. A cellular service provider launched a new calling plan that provided low off-peak rates for new subscribers. As hoped, the campaign resulted in a huge increase in off-peak airtime minutes. This should have translated into a corresponding increase in revenue. Strangely however, the company experienced a net revenue loss. After analyzing further, the company realized that many of the customers attracted to this offer had turned out to be those who were most likely to default on their bills. And the remainder consisted largely of existing subscribers who defected from the more expensive calling plans to enjoy the cheaper rates. Not only did the service provider cannibalize its more profitable business unit, it incurred additional collection expenses and lost revenues by attracting the wrong type of subscriber.

In this example, the campaign would have been much more successful if the company had done some up-front analysis of the type of customer it wanted to attract and targeted the campaign to this particular customer. Rather than undertaking a broad-based campaign, for instance, it could have started by targeting only those customers who have a good credit history. It could also have targeted prospects with a low turnover history (i.e., stayed with a service provider an average of 9 months or more), to increase the chances of winning over a long-term customer who would not defect the moment a competitor offered a slightly lower rate. Getting access to this type of profile information about your customers can be tedious but not difficult. You can consider using a knowledge warehouse to obtain this information and augment or cross-check the data with the profile information you receive through your loyalty programs. In fact, established telcos have huge volumes of data stored in their legacy systems, which they frequently tap into to gather customer intelligence. The problem is, this data can quickly become outdated and inconsistent unless the carrier invests in putting together a plan to gain market intelligence. This may well include tapping into a variety of data sources, such as market data, the carrier's own data, and competitive information. The objective of this is twofold: one, it helps you determine how you are performing and where you should be heading; and two, it helps you develop highly targeted calling campaigns that avoid the pitfalls we noted above.

Another component that makes it difficult to measure the quality of growth is the churn factor.

CHURN

Churn is the tendency for certain customers to shift rapidly from one service provider to another in response to moderately small changes in price or other incentives. Because of the fairly standardized offerings and resulting price corrosion in their markets, cellular and long-distance service providers tend to be especially vulnerable to churn. Every service provider has to contend with a certain percentage of churn, regardless of how well the service provider serves its customer base. What is an acceptable percentage? Churn rates tend to vary from one telecom segment to another. Based on averages in their particular market segments, most service providers factor a pre-set churn rate into their business and financial analysis. Moving above this rate can be a signal to start worrying.

Every service provider's goal is to minimize churn; or at any rate, try to hold on to the profitable customers who generate the bulk of the business. Some questions that can help you identify such customers are: What products and services are customers buying and in what combinations? Which are our most profitable products and why? Are our price and service incentives attracting and retaining our most profitable customers?

Once you have identified your profitable customer groups, your priority is to retain these customers. Here are some ways that you can go about creating stickiness:

• *One way of reducing churn is to try to match service offerings as closely as possible to the requirements of your desirable customer segments.* To help you segment customers based on their usage patterns and demographic profiles, you can utilize data warehousing tools. Several companies offer data warehousing applications that provide the knowledge to fine-tune service offerings so that they more closely match the requirements of a particular segment or demographic group of customers. Knowledge applications derive this data from an analysis of the data contained in the telco's call detail records (CDRs). These records are automatically created by the switch whenever someone lifts the receiver off the hook. They include detailed information about the call, such as time, duration, and called and calling party. CDRs are typically used for billing and customer profiling purposes. In addition to tapping CDRs, warehousing tools can also be set up to pull customer information from other sources, such as billing, customer care, call centers, and network switches.

- *Good customer care also plays a vital role in reducing churn.* Your profitable customers are likely to be more quality and service conscious, and will react favorably not only to customized service offerings, but also to the proactive customer care that accompanies these offerings. Providing more proactive customer care was discussed earlier.

- *Churn can also be reduced by offering usage-based incentive programs.* These encourage marginally profitable customers to stay with you longer than they would normally have. Of course, the longer a customer stays with you, the better your chance of instilling loyalty, and the greater the profit potential.

CREATE WELL-DEFINED MEASURES TO ASSESS AND ACHIEVE YOUR BUSINESS GOALS

This is a function that upper management and/or the CEO must lead, but involves heads of the different departments in an organization. You can create a more cohesive and hence defensible business by understanding and managing the interdependencies between the financial, customer, internal process, and people aspects of your business. Equally important, by doing this you will be able to define and measure key success criteria as an integrated "single" or "whole" enterprise and not as discrete business units.

There are five key organizations that must be in lockstep for a company to execute efficiently: business, finance, R&D, administration, and operations. Each of these organizations must share in the measures you define to assess and achieve your business goals. To begin with, it is essential that top management share the business plan at all levels within each organization: this boosts employee morale and assures buy-in on strategic direction. This also paves the way for the company to identify tangible ways to measure results and assign accountability for the results at all levels, including non-executives.

One way to create well-defined metrics is start at the top. Lay out your key business objectives. Clearly specify targets each objective: where you want to be and when. Then define 2–4 programs or initiatives to accomplish the targets for each objective. This is followed by a set of measures or metrics (how you will know you succeeded) for each initiative linked to that business objective.

Follow with regular (bi-weekly or monthly) operational review sessions where each department head has full accountability for reporting and defending the numbers. Issues can be confronted head on for the group to deal with collaboratively: there is generally no finger pointing, no defensiveness, and no place to hide.

As an example, let us assume one of your key business objectives is to improve customer perception of the value and quality of your service offerings. A key measure of this is your churn rate, which currently stands at 10 percent. Your target churn rate could be 6 percent, which would be achieved over a period of 6 months. Your game plan to accomplish this target consists of several programs or initiatives, namely 1) launching an end-user awareness campaign around the features and quality of the service combined with a promotional offer, 2) conducting sales and partner training to educate them on service attributes, and 3) expand the customer service program to proactively manage the process for tracking and resolving outstanding service complaints. Each program carries with it a metric to track its success rate. For example, the success rate for the customer service program could be an 80–90 percent customer satisfaction rating, with the metric being customer satisfaction rating. The success rate in the sales and partner training programs could be an increase in revenue from signing up new customers. Similarly, the success of end-user awareness campaigns could be measured by the customer response rates to the promotional offer as shown:

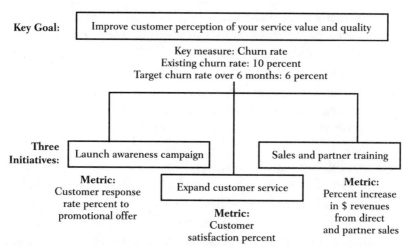

FIGURE 3.2   A framework for creating measures to achieve key results.

SUCCESS RATE	PERCENT ACHIEVEMENT*
1. Wild success	20+ percent response rate
2. Definite success	5–20 percent
3. Questionable success	2–5 percent
4. Definite failure	Below 2 percent

(*Note: these response rates are arbitrary. In general, response rates vary depending upon the unique characteristics of the promotional offer as well as industry norms, and therefore the response rate's success should be judged based both on the industry norm as well as the uniqueness of the promotional campaign).

Assigning metrics to each initiative enables management to get a read on where the results may be falling short and respond promptly. It also gives a more granular view by showing how much each initiative's success or failure rate is impacting the achievement of the target result, in this example the churn rate.

WORK TOWARDS FOSTERING LOYALTY

Because the costs of winning a new customer may be two or three times the cost of retaining the same customer, you must do everything possible to foster customer loyalty. In addition to developing the trust that creates loyalty through a series of service-based value exchanges, you can foster loyalty by offering incentive programs, such as frequent flier miles, rewards points, or free minutes. Until recently, the concept of loyalty programs was new to all but the airline industry, but today even grocery stores have loyalty programs. Many industries have followed suit, offering a range of programs from rewards points that can be redeemed for a variety of purchases to frequent flier miles to merchandise discounts, and so on. Loyalty programs are beginning to find their way into the telecom industry; for example, MCI Worldcom has teamed up with United Airlines to offer a frequent flier program to its customers.

Loyalty programs accomplish two key objectives: First, they attract customers to you for reasons other than the merits of your service. This can be key if you are entering a new market and have not yet established a reputation based on your products and services. If the program provides incentive enough, and you have not faltered in your customer relationship, customers will also keep coming back. Second, they can be a powerful information-gathering tool. Loyalty programs unknowingly force your customers to share with you vital information about their buying needs and preferences, whether through their travel plans, the types of magazines they subscribe to, or the type of stores they visit. In today's intensely competitive environment, your advantage lies in your ability to capture unique purchase information about your customers—information that is not accessible to your competition. Loyalty programs thus accomplish an important objective: they can educate you about the social and economic profiles of your customers, thus creating a powerful storehouse of knowledge that you can analyze and leverage to tailor your service offerings to their needs. You can also utilize this information to create targeted rather than broad-based marketing campaigns around new service offerings, such as those discussed in the section on tracking growth.

C H A P T E R

F O U R

PROFITABLE
MARKET ENTRY
STRATEGIES

Your market entry strategy drives your service strategy and lays the foundation for sustainable market share and profit growth. This chapter discusses the importance of creating defensible market entry strategies that will hold up in both good and bad market cycles, and lays out best practices for selecting your target market.

We all know the basic truth that entering a market demands careful planning and scrutiny, and that, depending on the experience of your people, your core competencies, product lines, available resources, and status (i.e., whether you are an established player or a new entrant), entering a particular market can be highly successful or not. What we often do not spend enough time understanding is which of these elements are most likely to influence profitability and success and which mistakes will surely result in bankruptcy. Here is an underlying wisdom that most young (and, dare I say, mature?) service providers fail to grasp: market share, early payback, and profit growth are achieved not only through efficient execution but also by execution in an environment where the company's capabilities are aligned with the chosen market.

The single most important element that influences profitability and market success is aligning your capabilities with your chosen market. Before entering a market, successful companies have already aligned their capabilities with the market and are thus able to quickly create and foster an environment for efficient execution.

To align your capabilities with your chosen market, you must both know your market and your own internal strengths and weaknesses—your core competencies. It sounds an awful lot like just plain common sense or a self-evident truth. How, you ask, could failing to do this result in such dire consequences as bankruptcy? Does this mean companies should never venture into potentially lucrative but unknown territories, where they have little or no knowledge of the market, even when they have access to capital and a management team that excels in execution? The answer is, in most industries, companies can do this and get away with it and sometimes succeed through sheer execution power. And even if the company realizes it has made a mistake and gone after the wrong market, the losses are usually not insurmountable. The company can still get out and recover. In telecom, however, it is a different ball game. The rules that apply in other industries don't apply here. The stakes are big. The up-front investment is huge. And the competition is brutal. Entering a telecommunications market, even a small niche, warrants a vast amount of preparation and planning. Once the investment is made and you're in, it is difficult to pull out without suffering the kind of multi-multimillion dollar losses that could crush you. In summary, you get one shot at picking your market. So, you had better get it right the first time.

Even when everything else is in place, this alignment of capabilities with your chosen market can mean the difference between success or failure all by itself. To illustrate, let's examine the outcomes for two young competitive service providers who, while they operated in different markets, had similar fundamentals and financing: Northpoint Communications and Everest Broadband. Both were founded in the late 1990s and both were well funded—if anything Northpoint was better funded than Everest, to the tune of $1.6 billion. Both set out to conquer their respective markets with an arsenal of services, marketing savvy, and top talent. Yet, a mere three years into their life, the outcomes for the two companies have been very different. Northpoint is bankrupt—and Everest is thriving so far (as much as can be expected in this volatile market). Northpoint made two fatal mistakes in its market entry strategy: first, it spent hundreds of millions of dollars to go after a national market where it already had two major competitors (Covad and Rhythms) and no hope of clear market ownership; and second, it went after a large market segment, the residential consumer, where it clearly lacked the core competency to execute.

Tempted by the lure of additional revenues and the possibility of filling underused business DSL lines, Northpoint severely underestimated both the brutal competition in the residential consumer segment, where the ILECs reign supreme, and the financial resources it would take to serve residential consumers. Northpoint's core competencies were not aligned with the market segment that it chose. By contrast, Everest used its money wisely and went after a niche market where it had the experience and resources to execute. Lessons were well learned by both companies on what works and what doesn't.

So, how do you go about selecting a market in which you can dominate? It requires a great deal of internal assessment and up-front market research, but the rewards are well worth it. There are several vital analyses you will need to perform, and these are discussed next.

ASSESS YOUR COMPANY'S CORE COMPETENCIES

This is an internal assessment that relates to the experience of your people; the types of customers your company has previously done business with or is currently engaged with; existing product lines, if any; and the knowledge base within your company. Understanding your core strengths and weaknesses helps you define the type of markets and services you are best equipped to go after, what your outsourcing needs are likely to be, and which partners exist to fill these needs in the chosen market. You can then begin to get a clearer picture of your target market profile.

A company's core competencies build from the experience of its people and the knowledge set that they bring to the table. Understanding your core competencies helps you evaluate your chances of success in a given market. For example, if you are planning to operate as a CLEC, you have most likely built an executive team from the ILEC, CLEC, or IXC space. They understand switched services, such as voice, very well. Having done it before, they can point out roadblocks and anticipate problems early on in the planning cycle. Delivering a data-centric service, like Web or application hosting, on the other hand, requires a very different skill set and experience—a core competency that your voice-oriented team and company as a whole may be lacking. Experts believe that making the assumption that the skills in one area can easily be spread to another is very dangerous. It isn't what you know that can hurt you. It's what you don't know and were not aware you needed to ask.

If you want to go for a broader voice and data service offering, that includes managed services for instance, you need to round out your team to include data experience or rely on your vendors or technical staff for advice and direction. The latter could well be a risky proposition. If you are unable to build in-house expertise, partnering with a managed services provider is the best approach.

On the flip side, if you are a managed services provider delivering fully managed services such as VPN or application and Web hosting, then your core competencies are on the data side and do not lend themselves well to delivering voice services. Most VPN and other managed service providers have successfully gotten around this dilemma by partnering with a facilities-based CLEC or ILEC and IXC for local and long distance voice services.

FIND A NICHE THAT YOU CAN OWN

With brutal competition from surviving CLECs and the well-entrenched ILECs, this may seem difficult to do, but it is achievable. Until you overcome this hurdle and identify a spot that you can own, you will be waging a "me too" war with the incumbents and other competitors. In a "me too" scenario you are forced to spend huge amounts of marketing dollars to out-sell the competition. In the end, unless a miracle occurs, the chances of your grabbing the lead spot are still minuscule. In fact, the lead spot is likely to be shared among several com-petitors, and your market position will be constantly threat-ened, leaving you with less time to focus on the customer and forcing you to engage in continuous irrational price wars. Moreover, it is very likely that you don't have the advertising muscle of the "big boys," so how are you going to position your-self for success?

The strategy to overcome this "crowded market" predica-ment is to initially stay away from markets where the com-petitive landscape is crowded. Identify a small or regional market segment where you are uniquely positioned to suc-ceed by virtue of your core competencies. Carve out this market niche for yourself and position yourself to own this niche. This creates significant barriers to entry for competi-tors and gives you a form of leadership or dominant position while you establish yourself and prepare for war with the competition that inevitably arises. In fact, niche-oriented service providers showed remarkable resilience during the market turmoil these past two years when compared to many larger, national service providers that were forced into bank-ruptcy. These national providers thought they would outsell the competition and spent millions in a mad rush to build national networks. According to Brad Baldwin, an analyst with International Data Corporation, "CLECs went into markets expecting to have total ownership of the market." I agree that certainly spelled disaster for national players such as Northpoint, but in fact this same strategy proved to be viable for niche players such as Everest Broadband and New Edge Networks.

So what is a *niche*, in telecom terms? Most likely it is an underserved market segment or group, either in the residential sector or small- and medium-business space. It is difficult to find large "underserved" segments. It is possible to serve niche markets on a national level, for instance small rural areas or Tier 3 cities throughout the U.S., but this is typically out of the realm of a small startup CLEC or service provider, because of the higher infrastructure expenses relative to the revenues. An exception to the rule, New Edge Networks, is an example of a CLEC that grew by targeting a national niche.

When you provide service to underserved groups, or when you provide services that aren't otherwise offered, you avoid pricing wars and thus set the stage for expansion into more competitive services. Plus, you'll increase your profits, margins, and attractiveness to investors. Because you understand your segment so well, your marketing efforts are very targeted, yielding lower cost of sales and far more predictable revenues.

THE CROSSOVER CHALLENGE— SCALING TO LARGER SEGMENTS

There is, however, a pitfall in adopting a niche strategy. For long-term growth you must select a niche segment that you can leverage to enter other related, larger segments. You also must identify the crossover point. Failure to cross over from a small niche to a larger market segment can be disastrous. Ricochet (Metricom) offers a good example of a company that was unable to cross over from a small niche to a larger one. Ricochet initially came out with a highly successful wireless-Web service targeted to a small, well-defined user segment. Unfortunately, the company did not have a crossover strategy and was not able to move outward to a larger segment. It subsequently went out of business when the entry-point niche could no longer profitably support the company's expansion.

And that is the caveat with niches: eventually you will maximize your market share potential and run out of room to expand. Unless you can successfully implement a crossover to a larger related segment, you are doomed to stay small, and shrinking profits will eventually force you to curtail expansion and be squeezed out by younger, more aggressive players.

DEFINE YOUR OUTSOURCING STRATEGY

What role will outsourcing play in entering the market? Which core competencies (that you lack, but need to be in business) can you offload to an outsource partner? What are your choices in partnering? The answers to these questions dictate your outsourcing strategy.

Aside from the skills that any startup, telecom or otherwise, requires, such as accounting, human resources, marketing and sales, finance, and vendor relationships; a service provider requires many specialized skills that are unique to telecom: customer support call center, provisioning, billing, and regulatory expertise are some of the key skills. Expecting to build an organization from scratch, with all the right personnel who have all the right skill sets, is unrealistic to say the least. Moreover, you are under pressure to keep costs low and improve time to market. In a situation like this, time to market can be significantly improved by outsourcing the non-core business tasks and even some of the core tasks. When managed properly, an outsource partner provides an immediate service that you may be unable to provide efficiently for another year or two. In that time, your churn rate and overall expense problems may not keep you in business.

In addition to outsourcing some of your in-house business functions, you may want to look at finding outsourcing partners for your service needs. If your strategy calls for being a Home Depot of applications, for instance, you will likely have to obtain services you can't provide with your own equipment. Some service providers have built their entire marketing strategy around outsourced services united on a single bill. Several studies show that consumers generally prefer to buy consolidated services and billing from a single provider. Certain applications, like prepaid calling cards and conference calling services, are available from a number of third-party providers and can be readily outsourced. Other applications, such as ASP-type business productivity applications and vertically integrated application packages, are also good candidates for partnering.

DEVELOP A VIABLE FINANCIAL MODEL AND BUSINESS CASE

Once you have identified a set of strategic niche markets where you are comfortable executing and have a reasonable chance of market ownership, you must analyze the financial viability of each niche market. For each market, you must determine that market's profit potential and business case or *return on investment* (ROI).

THE MARKET'S PROFIT POTENTIAL

Out of the markets you have identified, which market holds the most profit potential, both short term and long term? Put simply, you will want to focus on those markets and services where you have the greatest revenue potential, relative to the size of your investment. In sizing the profit potential of a market, a good yardstick you may want to use is the net revenue you will derive from your investment, relative to the size of the investment. If you have only a certain amount of money to invest, and your cost of doing business is mostly fixed, you want to put it into markets and services that generate the most profits by requiring low incremental investment to increase market penetration. To illustrate this point, let's analyze the profit potential of serving a multitenant unit (MTU) office building versus a group of single-dwelling offices.

An MTU building has a high concentration of businesses in a proximate location, generating many thousands of dollars per month in revenues for a service provider. Moreover, while initial take rates may be as low as 10 percent in a given building, the potential for generating additional revenue (by driving up the take rate) from the same investment is very high. This is because the building has already been wired for services. The high-speed connection from the public network enters the building at the basement, either through a DSLAM or an Ethernet switch, and is available to each business tenant on every floor through the building's wiring. Turning on new customers in this "smart" building now becomes not only cost effective but also focused, yielding higher returns from the initial investment. The service provider can leverage the tenants' similar demographic profiles and spending habits to create highly targeted marketing programs and partnerships. By contrast, acquiring new single-dwelling business customers with potentially different profiles will require additional, probably substantial, marketing and sales dollars, even when the single-dwelling businesses are located in close proximity to each other.

Deploying high capacity network facilities in the basement of an MTU building is also relatively inexpensive when compared to the cost of deploying facilities in the ILEC Central Office (CO) to serve a revenue-equivalent group of businesses. An MTU building basement colocation is a less costly alternative to the ILEC CO colocation, which is highly variable and can range in price from $35,000 to $250,000 (source: Yankee Group, 2000) for a preconditioned 10 × 10-foot cage. Given that, it is not surprising that several CLECs emerged to serve "underserved" multitenant office buildings. For the building-centric CLEC (BLEC), the multitenant office market offers a way to grow market share and revenue substantially with minimal capital outlay.

The inability to develop good financial or business models and accurately assess the profit potential of a given market can be perilous. Let's take mPower Communications for instance. mPower Communications is a good example of a service provider that was not able to accurately assess the profit potential of the voice over DSL (VoDSL) market. After moving quickly into several new markets with its innovative VoDSL offering, the company ended up losing millions of dollars when profits did not materialize as expected. Eventually mPower had to pull out of several of these unprofitable markets and refocus on its profitable circuit-switched voice network in an effort to stay alive.

The lesson learned here is that while new technology offers opportunities for innovative services, sound financial and business modeling is needed to validate the profit potential of the new markets created.

THE RETURN ON INVESTMENT (ROI)

A company's critical decision to enter a market should depend not only on the ability to align with the market and create market-based growth, but also on the nature and manner of costs that must be spent to enter the market. In other words, there must be a return on investment within a reasonable time period to justify the initial investment. You will want to carefully evaluate the profitability and ROI on your investments, both initial and incremental. When, and how, will positive ROI be reached? What variable and fixed costs will your company have to incur to reach a particular market? Will the payback period be a lengthy one or will payback be reached, say, within 18 months? What are your platform decisions going to be? Should you enter the market with a facilities-based approach or opt for resale? The answers to these questions can be obtained by building a financial model and business case analysis using a combination of service and network scenarios. Beware of faulty assumptions that can distort the ROI and payback picture.

When asked about the importance of ROI in determining profitability and payback, a senior executive from a survivor CLEC cautioned that ROI calculations can quickly become "garbage in, garbage out." According to this executive, unless you are absolutely sure about the allocation of overhead and costs, you could get a very different ROI picture. Incremental ROI, for example, only makes sense if you have a stable base to work from. In general, for ROI to provide a meaningful payback scenario, you must understand and itemize *all* your costs, including the recurring and nonrecurring charges, equipment maintenance charges, and ongoing operating costs that make up every dollar of investment for every revenue dollar you get back. Your ROI analysis should be able to accurately answer the basic question: If we put in a dollar today, then how much of this dollar are we going to get back, and when are we going to get it back? Experts agree that in calculating ROI, there is no magic formula. You have to be extremely careful about how to allocate the different system costs and the assumptions you make about potential revenues. The most important early step is to carefully think through the business case for providing your new service. Getting the business and financial planners involved as early as possible to form a model for the service puts them on par with the technical and strategic planners. That synchronization of the financial dimension with the strategic and technical dimensions can make the difference in whether a service succeeds or flops. Below are the recommended guidelines for building a viable financial model.

BUILDING A FINANCIAL MODEL

The objective of a financial model should be to estimate revenue, investments, operating cost, general administration cost, and taxes. Remember that the network is expected to generate revenue for the life of the product. Depreciation, operating cost, general administration cost, and taxes are deducted from the revenue stream to calculate payback. To judge the acceptability of the model, the cash flow is calculated by adding back the depreciation to the net income. The cash flow can be judged two ways:

- The *Net Present Value* (NPV) is today's value of the series of resulting cash flows, and not the investment, at an interest rate of x %. If the NPV is higher than the investment, the project should be judged acceptable.
- The *Internal Rate of Return* (IRR) is the interest rate received for an investment (a negative value) and income (resulting net cash flow) that occurred over the period. The IRR is the preferred method and most widely used by telecom providers, because it includes an analysis of the investment and resulting cash flow over the period.

According to industry experts, some of the factors that could seriously distort the IRR are:

- *Internal time intervals.* If the payback assumes immediate install implementation, yet, in reality, the timeline takes 60 days, the basic premise of the payback is flawed. This distorts the IRR.

- *Overestimate of usage.* This is one of the most common mistakes. It is better to underestimate and come in over, than overestimate and fall flat.
- *Customer penetration capability.* Marketing and sales estimates of the number of customers are often overstated and not well thought out.
- *Lack of follow-through.* Many projects are sold on their ROI, but seldom is there any accountability or accounting done on whether the project actually met the numbers. This creates an environment in which truth does not have to exist; the better salesperson can baffle the powers that be with tremendous revenue projections.
- *SG&A (sales, general and administrative costs).* Costs related to back office issues. McLeodUSA, a CLEC on our survivors list, believes SG&A is the most variable and most controlling factor in determining IRR.

Network costs can distort your ROI as well. Some network costs can be disproportionately high if you are not careful about allocating based on a realistic assessment of your deployment scenario. For example, the link cost should be calculated by taking the average cost per link over a network, not the link cost of a fully loaded shelf. Based on your deployment plans, you should calculate the costs of a realistic shelf topology rather than that of a fully loaded shelf.

Key variables that must be input into the financial model are:

- *Capital and non-recurring costs (NRC).* This includes networking equipment; backbone equipment; NOC equipment (such as network management systems and spares); setup costs, which could be central office installation or basement POP installation; and one-time training costs.
- *Depreciation.* This relates to how capital expenditures are treated for ROI purposes. Central office (CO) equipment is treated as an investment depreciated over the life of the product (typically three or five years). The subscriber (CPE)

equipment is generally treated as expenses in the first year, but can also be depreciated over twelve months.

- *Recurring operating costs.* These are all recurring costs except G&A. Recurring operating costs typically include billing; direct sales costs; subscriber installation and provisioning/configuration; total backbone recurring costs; maintenance; building rights of entry, if any; and incremental and other equipment cost allocations.

- *General and administrative (G&A).* These are recurring costs as well and typically include marketing and selling costs, advertising, general administrative costs, and back-office administrative costs. Note that the "S" in SG&A is missing because S is a part of the recurring operating costs above.

The first steps in building the financial model are to establish revenue benchmarks for your new service and set an expected IRR to strive for. Here, you are bound by certain parameters: revenues, recurring operating costs, and the IRR constant. For example, you may want to maintain a constant IRR of 20 percent. You will also want to recover your recurring operating costs. At the same time, pricing is somewhat inflexible because competitive and market considerations, not operating costs, will dictate the price you charge your customers. Therefore, you will need to manipulate recurring operating costs if you want to maintain a constant IRR, or change the IRR percentage number you are aiming for.

MANIPULATING RECURRING OPERATING COSTS

There are certain recurring operating costs that you have some control over adjusting. You can manipulate these to maximize your IRR:

- *Backbone (backhaul) costs.* Backbone recurring cost is an important cost is based on your costs for backhauling traffic to the regional switching center and backbone transmission facilities. Facilities-based carriers incur significantly lower backbone costs than carriers that have to build their own overlay networks on leased facilities.

• *Customer support call center costs.* This cost is incurred to provide telephone and technical support to customers each month at a fully loaded labor cost per hour. Most service providers peg customer support costs on the assumption that 20 percent of customers call for technical support each month and use 20 to 25 minutes of technical support time at a fully loaded cost of $90/hour. Most service providers factor costs somewhere between $6 and $8 per provisioned line per month, for voice and data.

• *Bandwidth leasing costs.* Oversubscription is commonly used to calculate bandwidth requirements. This assumes that not all lines will be busy at the same time, so that deployed switch ports can be less than the number of end-user connections. The subscriber concentration ratio you use influences how much bandwidth you need in the last mile. The lower the subscriber concentration ratio, the more the bandwidth requirements and higher the bandwidth leasing costs. By using higher subscriber concentration ratios, you can lower your bandwidth requirements. Oversubscribing, however, can result in performance bottlenecks and create unacceptable delays. Optimal bandwidth usage requires that you maintain a balance between the subscriber concentration ratio and performance problems resulting from oversubscription.

SAMPLE FINANCIAL MODEL FOR CONVERGED VOICE AND DATA IN MTU BUILDING

This sample model shows cash flow and payback analysis for a startup investment in a targeted metro area containing 50 MTU buildings per area, and a constant penetration (for the sake of simplicity) of 12 subscribers per building. Table 4.1 shows the startup costs associated with building space preparation, basement POP setup, and cash outlays to open a regional switching center and Network Operating Center (NOC). The BLEC has decided to invest in a packet voice switch to bypass the local loop and offer converged voice and data services over DSL. The BLEC is purchasing long distance voice from an IXC and reselling it to subscribers. Note that this model also demonstrates increased profitability by offering voice as a packet application in addition to data access. The network architecture for service delivery is shown below.

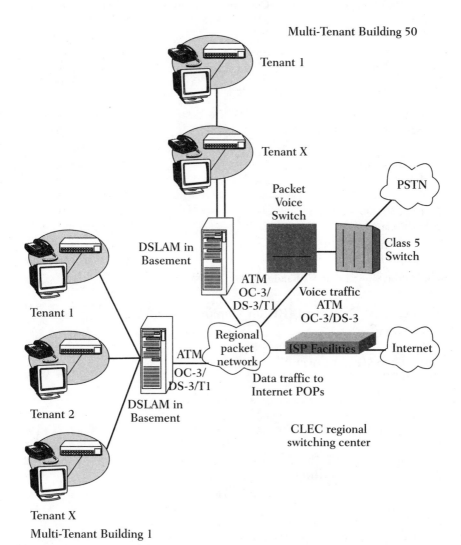

FIGURE 4.1 Converged network service delivery architecture for a group of MTU buildings.

TABLE 4.1   One-time CAPEX and Recurring Operating Costs for Deploying Converged Voice and Data DSL Services in an MTU

CAPITAL AND NRC	PROVISIONING	RECURRING
DSLAM: $7,000 per building	Customer acquisition $160 per subscriber	Backhaul from MTU $150–500 per month per building
CPE: $250–$1,050 per subscriber	Provisioning expense $150 per subscriber	Building ROE: 3–15% of gross revenue to property management
ATM switch/router: $135,000 per metro area		Customer support: $6–8 per subscriber per month
Packet voice switch: $250,000		Operational support: 6% of gross capital per year
NOC: $250,000 + 20,000 per metro area		Cost of long distance: $0.04 per minute
Setup and training (NRC): $3,000 per building		G & A: 15% of gross revenue
Total CAPEX (excluding CPEand NRC): $662,000		

TABLE 4.2   NPV and IRR for Deploying Converged Services in an MTU

CONVERGED SERVICES OVER DSL; NPV AND IRR YEAR END		YEAR 0	YEAR 1	YEAR 2	YEAR 3
Revenue (per month)	$560		$560	$560	$560
Cost of service (per month)	$368		$368	$368	$368
EBITDA (annual)	$192	($3,285)	$2,306	$2,306	$2,306
Net present values (NPVs)	15%	($3,285)	($1,113)	$403	$1,722
Internal rate of returns (IRRs)			–30%	26%	49%
Difference in NPVs		$1,347	$1,371	$1,544	$1,695

Table 4.2 shows both NPV and IRR for years 0 through 3 on a per-subscriber basis. If the cash flow is discounted at an annual rate of return of 15 percent (a quarterly rate of 3.6 percent), and the last year's cash flow continues for three more years, the NPV of the cash flow is $403 per subscriber in year 2 of operations. A more preferable method of judging cash flow,

the annual rate of return (IRR) of the cash flows is 26 percent at the end of year 2.

Table 4.3 breaks out the revenue, operating expenses, recurring and nonrecurring expenses, and capital expenses on a per-subscriber basis. It calculates a simple payback period on an absolute revenue and expenses basis, without taking into account start-up cash outlay. As additional subscribers are added, however, it is natural that you will see cost savings associated with spreading the variable overhead over a larger number of subscribers. Over time, however, revenues are likely to decline as well for the current service offering. So, you had best be prepared to make up for the revenue shortfall by constantly thinking of new ways to shore up revenues through innovative offerings.

TABLE 4.3 Payback per Subscriber for 1) Deploying Data and 2) Deploying Converged Voice and Data Services in an MTU Configuration

ITEM		DSL (DATA)	CONVERGED VOICE + DATA
1. Monthly revenue			
Local phone			$12.50
Intra-LATA toll	200 minutes		$5.00
Long-distance	500 minutes		$25.00
CLASS features	Included		$5.00
Per-line revenue			$47.50
Internet access	Up to 2.3 Mbps	$180.00	$180.00
Web hosting	Included		$0.00
10 email accounts	Included		$0.00
Per-subs. revenue	8 lines	$180.00	$560.00
2. Recurring costs	12 subscribers		
Lease ILEC/packet voice			$50.00
Backhaul from MTU	(4x T1 backhaul)	$33.33	$33.33
Intra-LATA toll cost	$0.015/minute		$24.00
Long-distance cost	$0.025/minute		$40.00
Building ROE	5%	$9.00	$28.00
Customer support		$6.00	$8.00
Operational support	6% of gross capital	$48.50	$100.50
Total RC		$96.83	$283.83

continued on next page

TABLE 4.3   Payback per Subscriber for 1) Deploying Data and 2) Deploying Converged Voice and Data Services in an MTU Configuration (Continued)

ITEM		DSL (DATA)	CONVERGED VOICE + DATA
3. Non-recurring costs			
Building space	$2,000	$166.67	$166.67
Backhaul set-up fee	$1,000	$83.33	$83.33
Provisioning		$110.00	$150.00
Customer acquisition		$80.00	$160.00
CPE equipment		$250.00	$1,050.00
Total NRC		$690.00	$1,610.00
4. CAPEX			
DSLAM equipment	$7,000	$583.33	$583.33
ATM switch and router	$135,000	$225.00	$225.00
Packet voice switch	$250,000	$0.00	$416.67
NOC (network management)	$270,000	$0.00	$450.00
Per-subs. cost		$803.33	$1,675.00
5. Payback			
Revenue		$180.00	$560.00
Recurring cost		$96.83	$283.83
G & A	15%	$27.00	$84.00
Gross monthly income		$56.17	$192.17
Non-recurring cost		$690.00	$1,610.00
CAPEX		$808.33	$1,675.00
Payback (months)		26.7	17.1

TABLE 4.4   Cash Flow Analysis for MTU

YEAR END	2001	2002	2003	2004	2005
COs	1	1	1	1	1
Subscribers	12	120	600	600	600
Total revenue	$80,640	$532,224	$3,628,800	$4,032,000	$4,032,000
Recurring costs	$40,872	$408,720	$1,684,800	$2,043,600	$2,043,600
Onetime NRC	$6,720	$67,200	$336,000	$0	$0
G&A expenses	$12,096	$53,222	$290,304	$241,920	$201,600
–As % of revenue	15.0%	10.0%	8.0%	6.0%	5.0%
EBITDA	$20,952	$3,082	$1,317,696	$1,746,480	$1,786,800
CPE deprec.	$12,600	$126,000	$630,000	$0	$0
CAPEX deprec.	$4,020	$40,200	$201,000	$196,980	$160,800
Net cash flow	$4,332	($163,118)	$486,696	$1,549,500	$1,626,000
Cumulative net cash flow	($657,668)	($820,786)	($334,090)	$1,215,410	$2,841,410

As the cash flow analysis shows us in Table 4.4, and Figure 4.2, positive cash flow starts being generated in the third year of operation. Cumulative net cash flow turns positive in the fourth year.

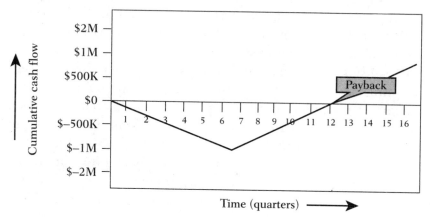

FIGURE 4.2   Payback using cumulative cash flow.

MAKING A BUILD, RESALE, OR LEASE DECISION

After careful market and financial analysis, you can decide whether it's more economical to build, resell, or lease network facilities to deliver services in your market. Financial and ROI constraints and time to market are all major considerations. Sometimes it is wise to resell services for quick market entry and build simultaneously for the long term. Heavy up-front fixed investment in facilities infrastructure has killed many CLECs, imposing heavy financial damage and lengthening time to market.

Successful service providers have chosen either approach based on their business and financial models, and there are arguments for both approaches. A number of CLECs, for example, argue against the scenario that if you build the network, customers will come. Most successful CLECs, however, believe that being facilities based is key to financial and operational success.

Owning the network certainly means higher profit margins, as well as more control over provisioning, service quality, and management. Mike Rouleau, Time Warner Telecom's senior vice president for marketing and business development, says that many large businesses use CLECs as secondary suppliers of data services for disaster recovery. If a CLEC is simply reselling ILEC network elements, it cannot provide the separate network that such customers want.

Resale, on the other hand, gives CLECs a foot in the door. By opting for resale, a service provider can enter a market without committing a lot of money up front, build a customer base, and then migrate customers onto its own facilities as they are constructed. This strategy holds that mixing wholesale ILEC capacity with home-grown capacity lets service providers make more economical use of funding. Ultimately, all say they eventually plan to run the bulk of their services using their own facilities. The ultimate reason to invest in facilities is higher profit margins. In this scenario, potential profit margins are very high, because the CLEC does not have to share revenues with an ILEC or another local loop provider. But, on a relative scale, capital expenditures are also high.

The Smart Build Approach

The best of both worlds is captured in an approach that is facilities based but leases facilities instead of paying for them up front. The provider leases the facilities for a monthly charge, converting to a fixed purchase when a revenue stream is assured and financial conditions stabilize. Leasing can be applied to both backbone facilities and network equipment. Recognizing the financial constraints of small carriers, several telecom equipment vendors are now offering their service provider customers flexible leasing plans as an alternative to outright purchase.

Allegiance Telecom, a facilities-based CLEC that provides local, long distance, and data services to small- and medium-sized business customers, entered the CLEC market using the "smart-build" approach. The company initially leased fiber optic line capacity or purchased dark fiber from other service providers instead of building its own fiber network in each of its markets.

Allegiance Telecom's market entry strategy to lease unbundled network elements (UNEs) and fiber capacity enabled the company to incur lower up-front capital costs and financial obligations. This "smart-build" strategy allowed Allegiance to establish a significant customer base in its markets before it built out its own metropolitan fiber transport facilities. Using this approach, Allegiance has succeeding in showing a fully funded business plan and does not have the same debt levels as many of its CLEC peers. At the same time, its interest expense is much lower than that of its peers in the CLEC industry.

Of course, not all successful service providers have chosen to lease network and backbone facilities. Time Warner Telecom is building a fiber optic network over which it has run all its customers from the beginning, with 24 markets already operational and another five planned this year. But Time Warner Telecom also places emphasis on being fully funded.

In the BLEC space, the term "facilities" becomes more nebulous. BLECs are required to provide the necessary infrastructure within the building, particularly equipment in the basement. Some BLECs, like Everest Broadband, have chosen to be partially facilities based, owning and operating the equipment in the basement, and providing their own billing systems and NOC, but leasing the IP network and ILEC services. Having a lightweight but strategically planned network infrastructure is key for all levels of provider.

MARKET ENTRY CONSIDERATIONS FOR ASPS

Application service providers (ASPs) comprise a specialty or boutique group of service providers. With the economic downturn, businesses are coming under increasing pressure to cut IT spending, including software applications and associated hardware. I believe that there will be increasing financial incentive for small, medium, and large enterprises to rent business applications rather than buy and maintain them in house, thus fostering a revival of the struggling ASP industry.

The complexity of the business model, however, poses challenges to ASPs that plan to enter the application hosting space. Which segment or segments to enter? What applications to host? Planning to enter the Web and application hosting business requires just as much up-front planning as for a "regular" CLEC. If you are planning to be an ASP, you must make the following additional assessments before entering this market:

- *Determine segment focus.* The small and medium business segment consisting of 25 to 500 employees is a lucrative target for most ASP business and productivity applications. Larger enterprises (500+ employees) have, in the past, been predisposed to buying rather than renting business-critical applications such as sales automation, supply chain management, accounting and payroll. With spending pressures on the increase in these businesses, however, you can now also consider targeting larger enterprises that were previously not candidates for renting applications. Vertical segments, such as high-tech firms, retail, professional services, and accounting firms, are also great candidates for ASP services. To determine where your focus should be, you will need to examine your core competencies, the profit potential in the different segments, and your chances of owning that segment.

- *Choose applications.* Target market segments vary in the types of applications they adopt. E-commerce is a popular hosted application, as are Internet call center, e-mail, and business process applications such as payroll, accounting, and customer relationship management (sales and marketing automation, customer support automation). Interestingly, you can add security and storage to this list. You may be tempted to offer a portfolio consisting of a wide range of services from basic to complex, but, as with any telecom service, focus is desirable at the time of market entry, particularly if you are planning to cater to a vertical or niche market segment.

One potential entry strategy for an ASP is to focus on simpler communication and collaboration applications rather than CRM or ERP. These applications typically do not require integration or customization, are easy to deploy, and have broad-based horizontal appeal among small businesses. iBasis, for example, is an ASP that offers hosted unified messaging applications. Microsoft, Oracle, and IBM offer simple e-business applications for hosting. Such applications require limited integration and customization and can be deployed fairly easily.

Another viable strategy, particularly for segment players, is to focus on building vertical baskets of applications that contain a mix of horizontal and segment-specific applications. A vertical application portfolio can be a profitable strategy for ASPs that want to partner with carriers to serve multitenant building segments, such as hotels, campuses, and small business tenants.

· *Choose whether to partner or build.* You will almost certainly want to partner with a software vendor for your application needs. British Telecom's Ignite, for example, is based on a partnership between BT and Oracle. To enter the market quickly, consider also partnering with a systems integrator that can provide integration and customization expertise. There are many examples of such partnerships in place. Qwest's Cyber.Solutions is a joint venture between Qwest and KPMG. Cyber.Solutions delivers infrastructure, software, application development and implementation, maintenance, and enhancements in the package. Sprint's ASP offerings are also based on a partnership with Deloitte Consulting.

Of course, you can choose to be a "pure play" ASP and build rather than partner. USi, for instance, is a facility-based ASP that manages all components of the ASP value chain, including application infrastructure. It has accomplished this full range of capabilities by building and acquiring application and system integration expertise. Other examples of pure play ASPs are Corio, ASP-One, Webex, and Interliant.

One thing worth noting is that the ASP market is consolidating and evolving, and customer requirements are also changing. The rental model is becoming more flexible, in pricing and delivery, to accommodate changes in customers' buying behaviors. Depending on whom you believe, projections call for this market to be several billion dollars by the year 2004. The three major impediments to widespread adoption of ASP applications today all have to do with support and reliability: the low-grade service level agreements (SLAs) that ASPs are handing out do not offer the level of guarantees required by business critical applications; the number of ASPs that are required to fill a single customer's IT needs is too many; and the contractual and support hassles businesses encounter when dealing with multiple ASPs are seen by customers as roadblocks.

Too many ASPs have entered the market, and most of them are application developers not equipped to offer a full-fledged ASP service. The natural forces of demand and supply will solve this problem. At any rate, because of the ease associated with renting, as SLAs become more acceptable, ASPs start providing more segment-tailored applications, and the ASP business model becomes clearer, it is almost certain that businesses will want to rent rather than buy all their applications.

Because of their knowledge of the broadband network and service delivery infrastructure, carriers are well equipped to enter this business, through partnering with an ASP or a software vendor such as Oracle or Microsoft. This is a good time to start thinking about getting into this business.

EXAMPLES OF NICHE MARKETING STRATEGIES

Below are two niche-marketing strategies that resulted in market ownership for these companies. In both cases, these companies aligned their core competencies with their target markets and identified initial niche segments that they later leveraged to expand into bigger segments.

A BUILDING-CENTRIC SERVICES NICHE: EVEREST BROADBAND

This case study focuses on one innovative niche player that has created a building-centric model for marketing and provisioning voice and data services to small businesses. Everest Broadband, a small service provider based out of New Jersey, has carved out a profitable niche for itself on the East Coast. The company figured out that a certain slice of the multitenant commercial building segment was underserved by existing carriers and built its service strategy on the premise that once it offered services at a fair price, this segment would become a loyal customer base from which to build. Everest continues to increase its subscriber base and portfolio of services by affiliating with quality providers of value-added services.

According to company executives, the company started out with the recognition that it would have a services niche. Therefore it didn't launch a large capital build-out, but rather optimized its investment in each building it serves. For initial deployment, only the work necessary to deliver services to the customers was done inside the building and at the POP. The company realized that it obviously needed transport services, but focused on investing only in the service opportunities, not the transport. Accordingly, Everest made the decision to not own a Class 5 switch, but instead operate as a "partial" facilities-based provider. Inside the building, Everest has its own facilities, along with facilities provided by the property manager. Outside the building, to the point of presence (POP), Everest leases transport facilities from the carrier. At the POP, the company has installed systems that perform aggregation and management of traffic.

Once Everest decided on its niche services approach, its next challenge was to further narrow its selected customer base—high-rise buildings—for maximum growth potential. Initially, the company planned to target both residential and office high-rise buildings. After a more thorough evaluation of the pros and cons of addressing both types of high-rises, however, it became apparent that Everest couldn't afford to focus its limited resources on what looked like two fundamentally very different marketplaces. The decision had its roots not in the service offering constraints, but in the type of customer care the residential market commanded and the lower revenue potential per unit. The residential customer requires a completely different approach from a customer care point of view: 24/7, versus a small business customer that typically requires 12/5. (Since most businesses work 12/5, the majority of their staff can be organized to support that requirement.) Servicing the residential market would thus have required a significantly greater investment in customer service and help desk staff. Moreover, a commercial building has potentially larger customers, therefore more revenue potential per unit. Founder and CTO Rashmi Doshi believes that, at some point, the residential market will become more attractive for Everest. In the near term, however, it poses a far higher risk than the company is willing to take on, and may hinder the profitability that the company is trying to achieve.

Everest then went one step further in selecting the target services and customers. In this area, Doshi observes: "A lot of it had to do with the experience of the people who came to work with us. The original team had worked in similar areas. After that, in terms of selecting the target services, a couple of things were key for us." Those two key tactics were:

- *Negotiating contracts with the building owners.* Part of the methodology here was that Everest would get a portfolio of the owners' buildings in order to offer services to those buildings. The portfolio typically includes buildings that are all ranges of sizes and have all ranges of tenants. Everest then assessed the revenue opportunity for a building based on the size of the customer (tenant). Everest looked for a category that fit the traditional definition of a small business—a company with less than 50 employees. Some buildings, for instance, are occupied by larger businesses, with several hundred employees, which occupy several floors of the building.
- *Identifying underserved customers that have a spending budget of $500 to $2,000 per month for broadband voice and data services.* These customers are ripe for broadband services. Everest found that customers with budgets under $2,000 to $2,500 per month don't get any special attention from incumbent carriers, whose account teams tend to focus on bigger customers spending between $10,000 and $15,000 a

month. Customers below this range are not big enough to justify special attention from the ILECs, and typically get a solicitation call only once every six months from the ILEC. Everest decided to focus their marketing efforts on this underserved customer segment.

On the question of marketing budgets and spending for customer acquisition, company executives say: "It's really a question of how you translate the marketing budget, not so much that we don't have a budget. First, we don't really need to advertise—we know exactly who our prospective customers are and where to find them. We don't have to put advertisements on subways and buses, only in the building, which is very focused advertising. Instead of media advertising, we use signage in buildings, and host lobby events such as coffee receptions, at which we introduce ourselves and our services to prospective customers. Property managers usually give us a list of who is in the building as well, and we do internal mailings to every tenant, often with a small giveaway or gift."

"It's very, very targeted marketing," Doshi observes. "It translates our acquisition costs from a broad all-over-the-board brush to a more focused investment per unit. It would not benefit us to spend any money on mass marketing."

Once the company decided on its niche strategy, the investment followed. Everest thus avoided the costly mistake of over-investing in network build-out and capacity, and then rushing to fill excess copper lines with unprofitable customers. Following its successful initial deployment in business multi-tenant buildings, Everest has since expanded its market segment to include hotel properties.

POOR MARKET ENTRY STRATEGY: NORTHPOINT

Conversely, Northpoint is an example of a company that elected to go with a broad market entry strategy. The company invested in massive up-front network build-out to serve the national small- and medium-business market. When customer take rates and revenues did not materialize as expected, Northpoint made the fatal decision to go after the residential market to fill underutilized capacity, thus further eroding the profit base it needed to survive.

An industry observer notes: "One problem was that Northpoint underestimated how fundamentally different the two markets were and how difficult it would be to serve both efficiently. As an example, a simple thing, but very significant from a business point of view, is that business customers are there mostly during business hours. For residential customers, however, you have to build a very significant customer care staff for after-hours service, because residential customers can call you at any time if they experience a problem or require service. If you have to do an installation, typically, they prefer that you do it after hours or over weekends, whereas with business customers you can do an installation during normal business hours. For Northpoint, this magnified the cost associated with building out of services, pushing back profitability, and eventually resulting in the company's downfall."

A "SMALL COMMUNITIES"-BASED NATIONAL NICHE: NEW EDGE NETWORKS

One broadband DSL provider selected a niche market strategy where it stood a better than fair chance of beating the competition but ended up doing even better. In contrast to regional niche plays, New Edge Networks has implemented a national "small-market" approach—focusing on small cities and towns and semi-rural areas across the U.S. The company's target market consists of one-CO towns that have a population of less than 50,000. The company offers broadband access services mostly to small businesses in these towns.

In many of these markets, not even the local ILEC has competitive DSL offerings, and where the ILEC does offer service, it is typically more focused on the residential market. The obvious negative in pursuing a small market approach is also a plus in that it keeps the competition away—prospective customers are simply thinner on the ground in small communities than they are in big cities.

New Edge's niche strategy is unique in that it is not regional. The company has instead defined a "national" niche. To execute such a strategy, the company had to install more CO infrastructure and spend more on connectivity than most CLECs to reach the same size customer base. But there is another side to the equation, according to company executives. "We wanted to go into markets where others weren't," says Dan Moffet, chief executive of New Edge. "In metro markets, you'd typically find Covad, NorthPoint, and Rhythms, probably others too, plus the ILEC, all in the same CO. In most of our markets, it's just us and the ILEC." It's clear now, with the demise of NorthPoint and with Rhythms and Covad also facing a host of troubles, just how cut-throat competition has been in those larger markets.

Smaller markets can be tough, too. While they may not be attractive to deep-pocketed national players, they can attract smaller regional providers. In the case of New Edge, the company initially competed with a few regional providers—Jato Communications, Vectric Communications, and ConnectSouth Communications—all of whom have since gone under. New Edge believes these companies were victims of the free-flow funding era, blowing all their capital up front and leaving no money for marketing through the tough times. New Edge executives have prior experience with marketing services to small communities, and this has enabled them to fight off the major national players that occasionally venture within the company's footprint.

The company also correctly forecast that demand for broadband access in small communities would be robust. As the CEO put it, "People in these markets want the services just as badly as people in metro markets—maybe more so." For many people in small markets, living there is a work- and life-style choice. But to get those benefits, they sacrifice easy access to goods and services, entertainment, and customers. Broadband has opened up a whole new window on the world for these communities.

SALES STRATEGY. New Edge adopted a pragmatic wholesale–retail business model. The company's preference is always to go into a new market with a good, strong local ISP partner. Relationships count in small towns. "You want the people who go for coffee with prospective customers or sit with them on local Chambers of Commerce," explains the company CEO. "People well entrenched in the community."

But where it couldn't find a strong and willing partner—some communities didn't even have local dial-up service when New Edge came to town—it had to be able to offer ISP services itself.

New Edge executives also understood the need for "guerrilla tactics" as opposed to "carpet bombing" when marketing in smaller communities. In metro markets or big cities, you typically have to orchestrate an expensive media blitz to capture attention. In small cities, the marketing is specific and targeted with much less reliance on media.

Marketing in small communities is usually done more through lists and direct mail and inside (telephone) sales. And it relies heavily on local knowledge obtained mainly from the company's partners. Local partners always help design the campaigns, although New Edge uses its own templates for small-town marketing drives. Local partners also share marketing costs, usually splitting them 50-50 with New Edge.

"They've got to have some skin in the game," Moffat says. But as relatively successful as New Edge has been with this fairly unique strategy on the broadband DSL side, it likely would not have survived if it had relied entirely on that business. Guerrilla marketing tactics to small communities can be replicated on a national level as New Edge has done. A potential drawback to entering a national niche—and one that keeps many smaller CLECs away—is the dilemma of investing in a national infrastructure without the high revenues to justify it. For example, one of the requirements of a national small-market strategy is that you need a fiber ring to backhaul traffic from all those little markets to the Internet. In the case of New Edge, this was not an incremental expense, because the company already had in place a backbone network with eighteen regional access points, which it uses to offer VPN and related services to regional and national enterprise customers.

The drying up of capital markets, although making it difficult for New Edge to implement a wider rollout in the near term, had another important and positive impact, making it difficult or impossible for any new competitor to enter its markets. And that means companies such as New Edge have, in effect, maintained a competition-free zone. When capital markets do loosen up again, the positions of these companies are likely to be virtually unassailable.

BACK TO BASICS: A TOOL FOR CONSTRUCTING YOUR MARKET ENTRY STRATEGY

We move now to a focus on using well-honed methodologies versus practical planning guides. Companies such as Everest may ultimately succeed by carefully charting their course through the market's choppy waters. Most new CLECs, however, were caught unprepared, as we have seen. If you are a new CLEC, this section and the following chapters provide critical starting-point insights for you. If, on the other hand, you are an existing carrier with a modestly successful service strategy, these sections can provide a blueprint against which you can measure and validate your process for defining your market and service strategy. No matter which, TeleChoice has developed a basic, "do-able" process for constructing your market entry strategy and laying a solid foundation for your telecommunications service. The top half of the diagram in Figure 4.3 presents the critical strategic work that feeds into developing a new service. These areas are briefly outlined below. The lower half of this diagram represents the service definition component that is described in the following chapter. The most important component in this process is "goal identification." The first goal is the business's goal (for example, "we are the industry's best at...."). Other goals may relate to market-specific goals, competitive goals, customer behavior goals, and so forth. The first and most fundamental goal, however, relates to the purpose of the business and its deepest competencies; it remains the most critical component.

In defining your market entry strategy, you must examine the following areas in detail:

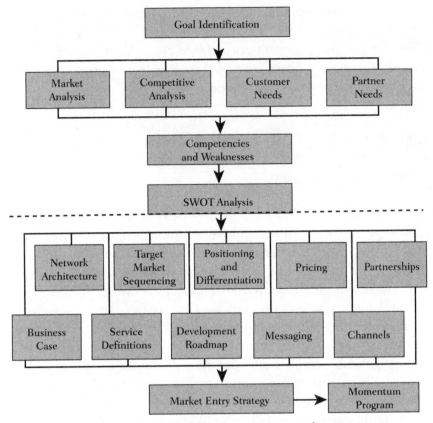

FIGURE 4.3   A basic methodology for determining a market entry strategy.

- *Goal identification.* What are your company's goals in launching the new service? For example, do you want to become a full-service provider, respond to a competitive threat, or become known as an industry leader? What are the specific goals of the service? Is it increasing market share, selling a certain number of units every month, or maybe increasing revenues per customer?

- *Market analysis.* What are the market drivers and trends that indicate the need for a new solution? What are the gaps in the current solutions in your target service market?

- *Competitive analysis.* Where are your main competitors in the target service market space? Against what services will yours be competing? What are their relative strengths and weaknesses? What opportunities can you take advantage of? What are the risks to your company?

- *Customer needs.* What customer segments have the most need for a new solution such as yours? What requirements do customers have today that are not being met by current solutions and their providers? What expectations do customers have? What are their main hot buttons?

- *Partner needs.* If you have close service partners, what opportunities do you have to fill their needs with the new service?

- *Core competencies and weaknesses.* What are the strengths of your company; what is it known for doing well? What are the specific weaknesses that you must work around in offering a new telecommunications service? For example, are you strong in offering data services, but have little telephony or voice expertise? What workarounds will you implement, and what will be the plan for arriving at an ideal situation?

- *SWOT analysis.* What are the strengths, weaknesses, opportunities, and threats (SWOTs) that your company is facing with the new service launch? What are the risks and implications of launching your service now or later? How will your corporate positioning be affected?

These are the main areas that you will need to examine and keep in mind when developing your service and market strategy. Because every service provider is different, there is no shortcut solution that will provide a competitive and differentiated service tailored to your situation. While this is a lot of hard work, you will find that developing a market entry strategy is very worthwhile and that the stress on your organization, once the service is launched, will be minimized if the strategy is well understood beforehand.

BUILDING AN EFFECTIVE SERVICE STRATEGY

This chapter provides the framework for creating an effective broadband service strategy. The service strategy culminates in a comprehensive service definition and rollout plan. The ultimate objective is to launch the service successfully.

There are three key areas you must focus on when creating your service strategy:

· Select the underlying transport technology
· Identify the type of services you want to offer
· Figure out how much to charge for your services

Partners play a key role as well. Some services can only be offered via partners. This is such a big topic that it is covered in Chapter 6. We begin this chapter with a discussion of the transport and access technologies to which you must have access to deliver your service to customers.

TECHNOLOGY'S ROLE IN YOUR SERVICE STRATEGY

You must formulate a technology strategy for delivering services. Your technology strategy could well be to take an agnostic approach to technology. In fact, many successful companies have not built their business models around the latest and greatest technologies. For example, the move to packet-switched networking certainly promises the ability to run more traffic in more ways over the same conduit, but a number of survivor CLECs say a company operating a circuit-switched network efficiently can do just fine. Several service providers say that they are operating a circuit-switched network and plan to migrate to packet switching, but these companies' business models do not depend on it.

In fact, more important than having the latest whiz-bang technology is acquiring customers and delivering the service they require. "People forget that telecommunications is a service, not a technology," says the chief executive of a competitive carrier. "You don't care what type of equipment your call goes on, you don't care whose backbone you're on. You want reliable phone service at a low price."

Customers don't care about the technology, only the service—a concept hard to believe in this technology-centric era. An industry insider has this advice to offer: "Before you leap at the next hot technology available from the laboratories of the telecom gurus, size up your market and know that the great service doesn't come from a software upgrade or new piece of machinery. It comes from a company-wide commitment to be a unified, no-excuse service provider to your current customers while consistently upgrading all departments to meet the market demands of the future."

RECOMMENDED STRATEGIES FOR CHOOSING A BROADBAND TECHNOLOGY

Still, your service needs a transport technology to travel on. Whichever broadband medium you choose to deliver your services, it will impact your financial return and the type, as well as speed, of services offered. Here are some strategies experts recommend when choosing a broadband access technology:

· *Adopt a technology-neutral approach.* Not all access technologies are available in all geographical areas, so you may limit your market coverage if your service strategy is too dependent on a particular transport or broadband technology, (such as DSL). One way to overcome this limitation, and one that several service providers are choosing to do, is adopt a "technology-neutral" approach. Your company can deliver services based on profitability and market acceptance, rather than the latest and greatest technology. By doing this, you can realize broader coverage and higher penetration potential within your chosen market. Moreover, even the latest technologies, such as Ethernet or optical, are not suitable for all applications. In conclusion, most effective market penetration is best realized by adopting a technology- neutral approach.

In the multitenant building space, it makes even more sense to adopt a technology-neutral approach. Buildings come in all shapes, ages, and sizes, and you will need to work with different infrastructures in different buildings if you want to penetrate more than a small percent of the market. CLECs like Urban Media, which centered their business models on a single technology—Ethernet—later admitted that it was not realistic to assume that the same technology can work and be profitable in

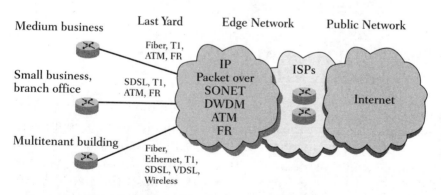

Being technology dependent can limit market share.

FIGURE 5.1   Some of the different transport and broadband technologies being deployed in the last mile in different segments of the small- and medium-sized business market.

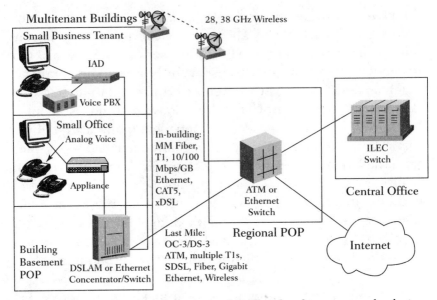

FIGURE 5.2   An example of the various broadband and transport technologies used in multitenant buildings and the last-mile network.

every type of multitenant building. Survivor CLECs have shied away from embracing a single technology, unless it happens to be available on the copper wiring that exists in 95 percent of buildings, and have focused instead on leveraging whatever technology is available to access their customers.

EVALUATE YOUR ALTERNATIVES AND COST IMPLICATIONS

Before you settle on a broadband technology to deliver your services, check out the alternatives available to you and analyze the potential headaches (by that I mean the financial and resource drain) associated with each one. For instance, you are looking to offer your customers 10/100Mbps Ethernet-to-WAN connectivity in a building. A closer check may reveal that the target building does not have (in fact, *most* buildings do not have) the Cat 5 cabling required to carry Ethernet. This means you will need to gain access to the building's internal riser system and lay the cables yourself—an expensive and daunting proposition! Rather than opting to go this route, you may want to look at other alternatives. For example, a smarter move could be to use VDSL—a hybrid technology that lets you take advantage of the existing copper wiring in the building (therefore low cost) and offers performance close to Ethernet speeds (keeps users happy and lets you charge the same as you would for Ethernet). DSL, more than any other technology, can make economic sense in older multitenant buildings equipped with copper or low-grade cable, because DSL operates over these wires and has the 4,000 feet or so reach necessary to cover almost any size building. Although SDSL does not offer the blazing speeds of Ethernet, VDSL is a comparable alternative.

As we discussed in Chapter 4, you must examine the costs associated with deploying services over alternative technologies in your target markets, and their implications to your payback period and return on investment (ROI) before you make your final service selection. Each new technology must also be judged for its total cost of ownership (TCO). TCO includes not only the purchase price of equipment, but also other variables associated with it, such as application support, training, maintenance, upgrades, and network administration. New technologies generally carry with them a higher TCO because of the lack of available expertise, software, and tools to support these technologies. Using new technology means that there is no history to warn us of potential deployment problems; a case in point is the provisioning issues associated with DSL. In general, the less complex the network infrastructure, the lower the TCO. Similarly, the use of existing infrastructure also lowers total cost of ownership.

CONSIDER YOUR MARKET POSITION

Your market strategy also dictates your technology decisions. For example, if your market strategy calls for serving smaller Tier 2 and Tier 3 cities, you will most likely have to consider copper- and coaxial-based transport solutions for the fractional T1 or DSL-based services you plan to offer in these markets. However, if you are planning to offer higher-bandwidth services, such as fractional OC3, in larger Tier 1 cities, then you will need access to fiber optic transport. If your target market is small businesses in multitenant buildings, then DSL, T1, or VDSL (because they run on the copper wire that is found in 95 percent of buildings) are great choices for your service portfolio—unless you are serving new buildings, which come equipped with Ethernet or multimode fiber.

DETERMINE BANDWIDTH REQUIREMENTS

When you are choosing a technology, a key issue is whether it will be able to support the bandwidth your applications require. For example, multimedia applications such as video on demand are bandwidth intensive, requiring as much as 2 Mbps for good quality transmission. You will want to research the approximate bandwidth that your services require, because this will influence your infrastructure requirements. The SLAs that you plan to offer will also shape your bandwidth and technology needs. Here are some guidelines that the DSL Forum published on bandwidth requirements for some commonly used business applications:

Remote access:	.014–6.0 megabits/sec
Computer telephony:	.128–1.5 megabits/sec
Videoconferencing:	.128–1.5 megabits/sec
Internet access:	.500–1.5 megabits/sec
Distance learning:	.500–6.0 megabits/sec
Web site hosting:	.500–6.0 megabits/sec

Source: DSL Forum, 2000.

CREATING A SERVICE PORTFOLIO: WHICH SERVICES AND HOW MANY?

You can be a basic services provider or an enhanced services provider. If you are just starting out as a CLEC, you have probably already figured out that basic services, like local and long distance voice and high-speed Internet access, will not be incentive enough for customers to sign up with you, or to switch to you from the incumbent service provider. Equally important to your bottom line, local and long distance voice are suffering from tremendous price erosion and are no longer the lucrative source of revenue they once were. These services are fast becoming low-margin commodities and could propel you into a price war that you have no hope of winning against the mighty incumbents. Moreover, how can you craft a superior value proposition based on offering only commodity services? The answer is, you cannot. You will need to expand into other areas. The question is, which and how many services should you include in your enhanced services portfolio? Should voice be a part of the portfolio, so that you can achieve competitive parity with the local provider? Should you focus on security services, Web hosting, e-commerce, or wholesale ASP applications?

For larger CLECs playing for big stakes, the key to success has been delivering as many services as possible, both because they offer differentiation and because the high cost of broadband deployment makes it difficult to realize profits without multiple revenue streams. Most next-generation, well-established CLECs and some aggressive ILECs—McLeodUSA, Time Warner Telecom, XO, Allegiance Telecom, and Qwest to name a few—offer a rich portfolio of services spanning voice to data to managed services.

Other service providers combine Web hosting, e-mail, and Internet access along with phone service, whether or not services are bundled in the sense that customers pay one bill. Mixing voice with data makes great economic sense, particularly if voice is treated as a packet application that shares the data network. Whereas analog voice consumes 64Kbps of bandwidth and thus uses the entire dedicated call connection, packet voice can be compressed to consume as little as 16Kbps, or even 8Kbps of broadband bandwidth. This allows multiple phone calls over a single connection and still leaves ample bandwidth for data applications. Mixing packet voice

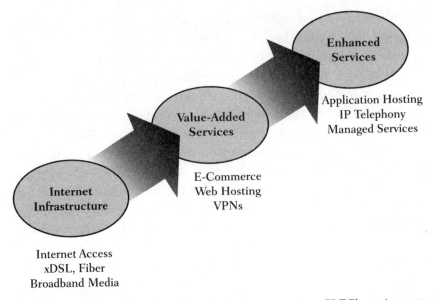

Enhanced
Services

Application Hosting
IP Telephony
Managed Services

Value-Added
Services

E-Commerce
Web Hosting
VPNs

Internet
Infrastructure

Internet Access
xDSL, Fiber
Broadband Media

FIGURE 5.3   Natural progression of telecom services in a CLEC's service portfolio.

with data thus assures service providers of both higher revenues and higher margins. Once the infrastructure is in place, it also becomes easier down the road to add advanced services such as VPN, firewall, and other high-margin services. More reasons for adding voice to the mix: it takes just as much effort to sell voice as to sell voice with Internet access, Web hosting, and firewall, and offering voice and data together has been shown to reduce churn. In conclusion, it makes sense for ILECs to continue to support analog voice because their infrastructure is already in place. For competitive carriers, however, investing in the deployment of packet voice, even only in the last mile as an initial service offering, is one way to rejuvenate voice as a new type of application, achieve differentiation from the incumbents, and keep revenues up.

If you are a brand-new CLEC, and only just creating a portfolio, then focus is desirable, and you may want to start with a vertical market basket of only two or three applications. You will need to investigate the newer broadband-enabled applications, especially those that deliver innovation along with convenience to small business customers. Below are some broadband-based applications that are in high demand and still far from being commodities—which means you can charge a nice premium for them. (Next-generation, softswitch-based services that are currently being developed are profiled in Chapter 9.

PACKET VOICE OVER BROADBAND

IP voice over broadband, or packet voice over broadband, is a last-mile packet voice application that you will want to consider because it not only allows you to address customer needs, but to establish competitive superiority over the incumbent provider in your market by offering multiple phone lines over the same broadband connection. Moreover, as we discussed, there is a strong economic case for adding voice to the mix: data is sexy and revenues are growing at a far faster rate than for voice, but voice revenue remains a significant portion of a customer's communications budget and is likely to stay that way for a few years to come. The Yankee Group believes that mixing voice and data is particularly attractive to service providers when the end-customer is a small- to medium-sized business: "If they are going after small- and medium-sized businesses, then the one way to make the customer more sticky, reduce churn, and get a higher share of wallet is to offer a voice/data bundle." Studies show that this business segment is highly receptive to bundled voice and data because it means one supplier, one bill, and simplicity along with discounts.

Note that, instead of opting for the more innovative packet voice service, if you have a strong or differentiated data offering but are looking to offer a one-stop-shop solution to customers, then you can simply lease analog voice from the ILEC in your area to complete your service offering. This is called logical bundling. If the services are bundled logically, they may appear on a single bill for the purpose of presenting a single interface to the customer, but they are physically deployed over separate networks, one for data and the other for voice. The data network is owned by the CLEC or ILEC. The transport technology may be fiber, DSL, wireless, T1, or even ISDN. The voice network is owned and operated by the ILEC. In this scenario, you lease the backbone transport from the backbone provider or ILEC for Internet access and data services. Local and long distance voice services are leased by you on a wholesale basis and resold to customers. This means that you lease the analog voice lines, or unbundled network elements (UNEs), from the ILEC for a negotiated monthly fee.

Logical bundling can make sense when you do not own your own facilities, and your market entry strategy calls for focusing on fast delivery of one-stop-shop service opportunities. You have decided to utilize partners to deliver your services, and your business model is based on deriving more revenue from a basket of services. Since packet voice is relatively new and still faces business and financial deployment challenges that may increase time to market, many CLECs are offering analog voice bundled with data service.

The more *integrated* service is packet voice over broadband. In this case, voice and data services are converged, which means the service provider is utilizing only one network, the broadband network, to carry both voice and data. Instead of traversing the traditional circuit-switched network, voice is packetized and shares the broadband data network. Analog voice is translated into IP packets at the customer premises using an integrated access device (IAD). A Voice over Broadband (VoB) gateway in the provider's POP or central office connects to the PSTN Class 5 switch using GR-303, V5.2, or TR-008 over a DS1 or STS-1 interface. The VoB gateway interfaces with the broadband switching and routing elements using a DS-3 or OC-3/ATM/STM-1 interface. The gateway performs the function of translating the packet voice back into traditional analog voice for delivery to the Class 5 switch. Packet voice is transparent to the customer in terms of quality and reliability, although it requires IAD at the customer premise and also has a lifeline support issue. Offering packet voice over broadband is appealing because it allows CLECs to bypass the ILEC's local loop and associated costs, not to mention the frustration of dealing with the phone giants.

Turning voice into a data application and making it a part of the data services package, makes it easier for you to put together a voice and data bundle. This approach usually offers higher profit margins, since you do not have to incur UNE leasing costs. Another significant advantage of implementing packet voice is that you can offer multiple phone lines to your customers over the same broadband connection. Because the costs are lower for you, you can opt to pass on some of these price savings to your customers, making packet voice more appealing to the customer than having to buy several individual, higher-priced analog POTS lines.

Several CLECs that are implementing this integrated services approach come to mind, but Allegiance Telecom is one that has done this successfully using T1 lines. Allegiance's service is profiled below.

INTEGRATED VOICE AND DATA OVER T1: ALLEGIANCE TELECOM. Allegiance Telecom's fastest growing products are its Total Communications Options and Integrated Access packages that provide customers with integrated voice and data services over a partial or full T1 access link. With this service, Allegiance places an IAD at the customer's premises that enables both voice and data to be transported over a single T1 facility. With integrated access services, Allegiance is able to generate higher gross profit margins because of the savings of not having to lease separate UNE voice and data facilities from the ILEC.

In early 2001, Allegiance expanded its addressable market for integrated services by lowering its requirement from 12 voice lines to eight voice lines. Since the inception of the eight-line minimum, Allegiance has experienced such tremendous demand that the company has not been able to maintain operational support for installing this product. This has resulted in a backlog of about 40,000 lines as of mid-July 2001.

Each integrated package can be configured in a variety of ways, but starts with a minimum of eight voice lines for each package. Each voice line in use requires a dedicated amount of bandwidth (64 Kbps each), which means that as more voice lines are activated the total amount of bandwidth allocated for data is diminished. Allegiance configures its Integrated Access and Total Communications Options package in several ways to match customer needs and future requirements. The minimum configuration for the packages includes eight voice channels with Internet access at 256 Kbps, which constitutes one-half of a T1. Maximum configuration of the package is with 20 voice lines installed, with 256 Kbps still maintained for Internet data traffic. The company provides partial T1, full T1, and multiple T1 bandwidth rates, using HDSL and HDSL-2 as the preferred method to provision the service.

Table 5.1 lists Allegiance's dedicated Internet data offerings.

TABLE 5.1   Data Offerings from Allegiance Telecom

DATA SPEED	PROVISIONING
256 Kbps	Partial T-1 line
512 Kbps	Partial T-1 line
768 Kbps	Partial T-1 line
1.54 Mbps	Full T-1 line
3.0 Mbps	Two Bonded T-1 lines
4.5 Mbps	Three Bonded T-1 lines
6 Mbps	Four Bonded T-1 lines
7.5 Mbps	Five Bonded T-1 lines
9 Mbps	Six Bonded T-1 lines
12 Mbps	Eight Bonded T-1 lines

Source: Allegiance Telecom/RHK.

Table 5.2 details the local voice services and features offered by Allegiance Telecom.

TABLE 5.2   Local Voice Service and Features Offered by Allegiance Telecom

LOCAL-SERVICE PRODUCTS	LOCAL-SERVICE FEATURES
Business lines	Call Forwarding
Analog PBX trunks	Automatic Callback
Digital PBX trunks	Call Transfer
ISDN PRI (Primary Rate Interface)	Call Waiting
Direct Inward Dialing (DID)	Call Hunting
Direct Outward Dialing (DOD)	Caller ID
	Distinctive Ring

continued on next page

TABLE 5.2   Local Voice Service and Features Offered by
Allegiance Telecom (Continued)

LOCAL-SERVICE PRODUCTS	LOCAL-SERVICE FEATURES
	Remote Access
	Three-Way Calling
	Toll Blocking
	Voice Mail

Source: Allegiance Telecom/RHK.

Allegiance provides long distance voice service through resell agreements that the company purchases at wholesale rates.

INTERNET TELEPHONY (VoIP)

Whereas packet voice over broadband packetizes voice in the last mile, IP telephony or VoIP uses IP throughout the public or private network. VoIP is the future solution for long distance voice and is discussed in greater detail in Chapter 9. International Data Corporation predicts that revenue from IP telephony services will reach $23.4 billion in 2005.

VoIP allows you to make a phone call using an Internet connection. Like packet voice over broadband, one VoIP connection can handle multiple (up to 24) POTS phone lines. Net2Phone is an example of a company providing VoIP to small businesses. The company charges a mere 2.9 cents/minute for calls within the U.S., although international calls are more expensive. Genuity is another company that has developed a VoIP offering, Black Rocket Voice, which is currently being given a trial by Cisco (for internal use) and Verizon Communications (for their services portfolio). Verizon owns an 8 percent stake in Genuity. Black Rocket Voice claims to integrate voice and data traffic onto a single, multiprotocol IP network infrastructure utilizing Genuity's Tier 1 IP backbone. Other VoIP providers include iBasis and ITXC. Most softswitch vendors are developing support for VoIP into their next-generation product offerings, which are still not ready for deployment in production networks.

In addition to enabling long distance calls over the Internet, another use for VoIP is hosted IP voice services and unified messaging in office buildings. By putting a multitenant IP PBX in the basement, you can create a shared-tenant hosted PBX model, for example. VoIP is finding its way into managed telephony services as well. iBasis, for example, provides hosted VoIP and VoFax services to international carriers, thus eliminating the implementation issues carriers may face. Yet another way VoIP is offered is through VoIP over virtual private networks (VPNs); this service is also called Voice over VPN.

A potential obstacle in implementing VoIP is that it requires the use of specialized IP telephones, which can be expensive because your customers will have to make existing phone equipment obsolete. You can get around this by providing upgrade incentives to customers, such as "giving away" a certain number of IP phones if the customer will purchase the service—similar to the cell phone model. Of course, the economics must be feasible. Another way to do it is offer to retrofit existing equipment. Net2Phone for example is making it easier for customers by offering hardware to retrofit both existing phones and PBX equipment.

Another potential obstacle in the adoption of VoIP has been the lack of quality of service (QoS) in the public network, and a corresponding lack of reliability. Even to the cost-sensitive small business, reliability is of paramount importance in accepting packet voice, whether it is VoIP or packet voice over broadband in the last mile. According to a service provider currently deploying voice over broadband, the technology itself doesn't appeal to customers: "What matters to the customer is a reliable high-speed data and voice bundle that keeps their costs low." Packet voice over broadband in the last mile has been tested and fits the bill, whereas VoIP is still in the early stages of deployment.

Not wanting to lose customers to next-generation carriers such as iBasis and ITXC, long distance incumbent carriers are investing in their own variations of VoIP, some offering VoIP over their private IP backbones to minimize QoS issues. For example, MCI Worldcom recently introduced its version of

VoIP, which carries IP voice over MCI's private backbone network. Because this is really VoIP over VPN, it does not have the reliability and quality issues that are associated with VoIP that uses the Internet backbone.

Depending on which type of voice you choose to bundle with the data services, packet or analog, offering voice could have a significant impact on the overall service portfolio, how it is positioned to potential customers, and when payback is achieved.

MESSAGING AND COLLABORATION

Messaging and collaboration services can be a lucrative source of revenue from small- and medium-sized businesses. These services usually find their way into most ASPs' portfolios, but can easily be a part of any service provider's offering. Most businesses utilize e-mail for corporate communications, so offering e-mail user management in buckets of 10, 25, or 50 users is a service you can bundle with Internet access. You can consider offering both domain and nondomain name e-mail. Document management, audio conferencing, Web conferencing, video conferencing, and unified messaging are other messaging service offerings. A popular Web conferencing platform is Genesys Meeting Center, offered by Genesys. Genesys Meeting Center is an integrated audio and Web conferencing platform that gives users voice and Web functionality from within a single interface for about $50 per month. They can collaborate remotely by scheduling and managing, online, both the audio and the Web component. Several other collaboration tools are available; another example is Microsoft's Net Meeting, which allows businesses to conduct virtual meetings. Other Web-based tools allow you to make presentations and share business applications in real time, without the need for costly travel or special facilities.

With messaging and collaboration services, businesses can conduct communications more efficiently. They can share presentations, documents, audio and video files with co-workers and partners, and control these communications via the Web. Overall, collaboration and messaging tools can be a profitable addition to your portfolio. You can partner with ASPs or the software vendors themselves to resell messaging and collaboration services.

VIDEO ON DEMAND

VDSL, cable, and other high-speed broadband media have made video and broadcasting applications possible. In addition to providing high-speed Internet access, you can deliver advanced services such as video on demand (VOD). VOD enables consumers to choose movies from an on-screen menu and control the sessions with VCR-like functions such as stop, fast-forward, and rewind. This is unlike traditional pay per view, where the consumer must watch the movie at a fixed time. Video on demand is being offered to the hotel industry by service providers such as On Command Corporation. Many small business customers are also interested in video on demand, because it offers inexpensive videoconferencing and broadcasting to their customers and partners, while providing a "big company" image.

UNIFIED MESSAGING

Long touted as the killer application, packet voice technology is only now beginning to enable unified messaging. Unified messaging integrates e-mail, voice mail, and fax onto a single platform (inbox) from which messages can be retrieved at any time from any type of device (e.g., PDA, PC, wire-line, or wireless telephone). Unified messaging also offers users the ability to route calls and messages depending on the customer's location and preferences. The idea is to provide quick, easy access to what had been separate messaging systems. To succeed, however, you will need to integrate legacy voice messaging systems with your e-mail, PC, and IP phone systems, while combining the different media types into a unified messaging environment. Alternatively, you can partner with a provider such as iBasis to offer the service. iBasis' VoCore is a hosted solution that integrates voice, fax, and e-mail messages in one unified mailbox accessible from any phone, browser, or standard e-mail client. Unified messaging is discussed in more detail in Chapter 9.

WEB HOSTING AND E-COMMERCE

A Web hosting service offers greater differentiation and higher margins than a basic Internet access service. Most small businesses do not have the in-house expertise to build or maintain a Web site, and more than 85 percent of small businesses that have a Web presence are outsourcing this function to their service provider.

Some of the services that you can offer small businesses include shared Web hosting with specific limits on bandwidth and capacity (disk space), e-commerce capabilities, and Web site design. E-commerce ranges from basic online storefronts, to more complex auction sites, to e-commerce. You can offer larger businesses dedicated Web hosting for more complex and custom Web sites. You can also consider offering a platform for premium Internet content, such as customized dynamic content distribution and video streaming.

Allegiance Telecom has perfected the Web hosting business by acquiring several companies to round out the capabilities of its Web hosting division. The company's Web hosting services include shared hosting, dedicated and managed hosting, and colocated hosting services, as well as a variety of Internet connectivity options.

Web hosting, when offered by itself, is typically a low-margin business. It becomes lucrative only at the higher end, when you can charge premium prices for customized hosting. This is because Web hosting can be expensive to implement, since data centers are expensive to set up and difficult to maintain, particularly for low-end hosting. The preferred way for a start-up to offer this service is through partnering with a Web hosting provider (such as Verio) to resell the service, or to bundle it with a lucrative offering such as voice or VPN.

TELEPHONY HOME DEPOT

You may want to go one step beyond voice and data bundling and expand to a business service area like ASP. Small- to medium-sized businesses represent a significant market opportunity for ASP rental applications because they do not usually have a wide portfolio of these applications available to them, because of the costs associated with purchasing and deploying these applications. You can leverage your proximity to the customer and your ownership of the last-mile network to function like a "Home Depot of Telephony," by providing a single point of contact to the customer for outsourced ASP rental services. To support this model, of course, you will need to partner with ASPs and ASP aggregators and integrate their services where billing, provisioning, and management enter the picture. Participating in the ASP industry is explored in greater detail in Chapter 6.

MANAGED SERVICES

A managed service is a complete turnkey service, that for a monthly fee, represents an outsourcing alternative for small businesses that do not have the in-house IT expertise to set up and host network and security systems. Instead of being hosted at the customer premise, the services are hosted remotely by the service provider. They are provisioned and remotely monitored 24/7 from a Network Operations Center (NOC).

Below is a list of services that you can offer small and medium businesses on a managed basis:

- Security (managed firewall)
- Branch-to-branch and remote access connectivity (managed VPN)
- Web hosting (managed Web hosting and e-commerce)
- E-mail (managed e-mail services)
- ASP applications: Enterprise Resource Planning (ERP), Customer Relationship Management (CRM) applications, and other business process applications
- Managed telephony services (such as hosted PBX, hosted call center), which are still in the very early stages of development. These services are discussed in Chapter 9.

MANAGED SECURITY SERVICES.   The most popular of all managed services today, managed security services are the outsourced management of enterprise security devices, systems, and processes. These devices and systems may or may not reside at the customer premise. In all cases, they are managed remotely.

The majority of broadband lines serve those least prepared to implement complex security measures: small businesses, telecommuters, and SOHO businesses. These businesses also do not have the IT resources to implement an adequate security solution inhouse. For this reason, managed security services are in high demand by small business and SOHO customers and are fast becoming the most lucrative of all services. In fact, IDC projects a $9 billion managed security services market, just waiting to be tapped.

A wide range of security services can be offered:

· Firewall management
· Intrusion detection
· Vulnerability assessment and testing
· Antivirus management
· Authentication and security
· Virtual private network (VPN)
· Public key infrastructure (PKI) certificates

Typically, to provide a managed security service, you will need to install the security appliance or device at the customer premise and manage it remotely from your NOC. You can also implement security functions in your network and eliminate installing the security appliance at the customer site. The pros of the latter approach are that you streamline expenses by potentially eliminating the truck roll, you do not have to worry about supporting multiple vendors' equipment or training your sales force in the many security appliances available, and you do not have to purchase the appliance, which can improve your ROI. The cons to the network-based approach are that customers generally want security to be located at their premises where they can see it and have some measure of control over it, and your network becomes more complex because you now have to implement and monitor high-availability systems in the network, such as load balancing, redundancy, and the like. Companies such as CoSine Communications, Nortel/Shasta, and others are promoting network-based security services. There are numerous vendors supplying CPE appliances and other security equipment for the customer premises approach, ranging from $199 at the low end to $20,000 for high-end products. If you are going to be implementing the CPE approach, then ease of use in configuring and some measure of self-provisioning should be key criteria when selecting an appliance. Because of the scores of CPE products available, you will need to select and support a handful based on the price and features needs of your target segment.

You can construct a security offering for small- and medium-sized business customers in many ways. Typically, very small businesses are not interested in very advanced firewall services or branch-to-branch VPN connectivity, because they usually do not have branch offices.

Here is a sample managed security portfolio with some sample pricing for a small business consisting of 80 employees and five regional offices in the U.S.:

- *Managed firewall.* ($199/month) Resides at headquarters, protects all offices with an ICSA-certified firewall, monitored 24/7 from the service provider NOC.
- *Managed anti-virus.* ($99/month) Anti-virus is installed and supported on all desktops.
- *Managed firewall with Web blocking.* ($249/month) Provides firewall protection and Web blocking services. Web blocking filters out undesirable content and is in high demand by schools and small- to medium-sized business customers.
- *Managed VPN.* ($349/month) Remote access VPN for telecommuters covers 5 to 25 simultaneous remote users; branch-to-branch IPSec 3DES VPN covers all five offices connected to headquarters.
- *Managed firewall with intrusion detection.* ($595/month) Protects the customer network with options for intrusion detection, load balancing, and high availability.

In this example, at the high end the average small business customer pays under $1,000 per month for a fully managed firewall with intrusion detection and VPN service. The service provider generates revenues of about $12,000 a year in addition to initial setup and installation; the latter is typically a break-even proposition. The service pricing in the example above does not include the price of the security appliance, which you can choose to pass on to the customer (bad choice), subsidize to the customer (customer pays 50 percent or less of the price as a "service setup" price), or incur the full costs yourself and take it as an expense for the first year of operations. Most security service contracts are renewable at the end of a year of service. You will want to make sure that you break even before the year is up, or consider negotiating a two-year contract, otherwise you may end up losing money.

In this service example, you can add other services and also mix and match services based on the preferences of your target customer base. Managed services can be delivered bundled or a la carte, or as private-label offerings. For quick market entry or to minimize up-front infrastructure costs, you may want to consider wholesaling from a third-party security provider. In this case, the third-party provider will expect to pocket about 70 percent of each monthly fee; the remainder falls through directly to your bottom line. You will incur costs of training during service launch, followed by sales and marketing, and the first line of customer support thereafter.

Alternatively, you could partner with a consulting organization that provides network security policy and planning services, and work directly with a firewall/VPN vendor to offer security services to your customers.

MANAGED NETWORK SERVICES. Managed network services (MNS) is a professional service offered by providers to providers. MNS focuses on the growing metro access market by breaking open the last-mile bottleneck and enabling customers to accelerate local metropolitan area network build initiatives. The build-out of metropolitan-area facilities is widely recognized as key to unleashing consumer and business demand for bandwidth-intensive applications like video and multimedia. Williams Communications recently launched a new Network Managed Services business to extend its professional, construction, and management services expertise to carriers, ISPs, and ASPs. The end-to-end services offered by Williams Communications include long-haul and metropolitan area network and facilities design; construction, design construction, and operation of POPs, data centers, and other facilities; optronics installation; and network management.

MANAGED DATABASE SERVICES. Managed database services (MDS) is another type of managed service that you can consider. Companies such as Oracle have tailored their databases to support remote hosting and management. If you offer this type of service to your customers, they can implement enterprise-level Web applications with a back-end database, without incurring the heavy up-front costs of licensing and deploying the software. Some of the components of a fully managed database service are having a DBA and highly trained database-proficient staff that can perform the following functions from the NOC:

· Procuring and maintaining the database license
· Installing and configuring a default database
· Proactively monitoring database processes and file system space
· Administering and maintaining a database security policy
· Performing online hot backups of the database

The database system and applications are hosted remotely on a dedicated platform. As with any other managed service, customers can shorten time to deployment for their mission-critical applications.

The above services are merely examples of what you can offer to your subscriber base. There are countless new types of services being created by application developers, and you will need to do your own research on what you need to offer to stay competitive.

FINAL THOUGHTS ON SELECTING A SERVICE PORTFOLIO

Being prepared to introduce new services means you must be able to stick not only to your budgets but to also alter your business plans. Telecommunications services follow the same marketing principles that are applied to the high-tech industry: profits from any single product can shrink over time as the technology matures and more competitors enter the market, so successful carriers must continuously examine the possibility of enhancing existing offerings or introducing new services to remain competitive.

David Ruberg, Chairman and CEO of Intermedia Communications, recently noted that service providers must keep their business plans fluid and adapt to rapid technology and market movements. "These business plans are not frozen in time. In 1990, when T1s were going for $2,000 or $3,000 a mile, and T3s were going for $12,000 per linear mile, you could afford to build yours because there was price stability and a great margin. Today, when a T1 goes for $250, you don't build them. So, what was appropriate in 1990 is not appropriate today. You have to adapt."

In the end, experts agree that the shakeout in the telecommunications industry will come down to which service provider can provide the most value to its customer base. Service providers that can find ways to differentiate commodity services and optimize the value of their existing networks by introducing multiple or enhanced services that leverage these networks will surely have the competitive edge.

In designing your rollout strategies for enhanced services, you must weigh all the variables, including market demand, demographics, and cost of service delivery. In the end, the highest profitability and financial rewards are attained by designing a service bundle that appeals to your target customer group.

THE NEXT VITAL STEP: CREATING A SERVICE DEFINITION

Once you have identified your technology and service portfolio, you must articulate these in written form so that you can explain your service plan to an internal audience and set the foundation for your service launch. The process for defining your service builds on examining several core offerings from the perspective of your company as well as the outside world. Service providers are often impatient to get to market and may perform only cursory analyses or skip them entirely to launch a service quickly. However, even though the process described here is thorough, it does not take an inordinate amount of time when resources are focused on each area.

Once translated from strategy into a documented service plan, the service definition becomes an internal reference document that describes your service to your own organization as well as to partners and suppliers. It is often shared with several groups within your company, although the marketing group is usually the owner of the document and maintains it as a living document. Any internal group that needs to understand the particulars of the service should have a copy. These may include sales, accounting, billing, marketing communications, customer support, sales support, and the like. This document is not marketing collateral, although it is often used as a source for creating collateral. A framework for creating a service definition is described below, starting with the service overview.

SERVICE OVERVIEW

Creating a service overview helps you describe the service and what the service supports at a high level:

- A short explanation of why your company is offering the service (e.g., to serve new customers, serve existing customers more cost effectively, to be a market leader for a certain target segment, etc.)
- Your target customers

- On-site installation of equipment
- The customer premise device (CPE appliance with integrated security features) installation and setup
- Voice service activation
- Six voice lines and associated usage (local, long distance, international)
- Calling features: conference calling, caller ID, voice mail
- High speed data connection for Internet access
- Basic remote Web hosting
- E-mail management service
- Remote access secure VPN and branch-to-branch secure VPN service
- Basic or premium SLA

SERVICE LEVEL AGREEMENTS

SLAs are especially important when defining a service to business customers. The more educated your customers are, the more likely they will demand some type of contractual commitment on the level and quality of service they are purchasing. An SLA is essentially a legal contract guaranteeing a quality of service. It is a way for customers to lock service providers into a particular performance level. SLAs are likely to become key factors in a customer's service decision as integrated voice and data services are adopted more and more by smaller and medium-sized businesses.

Even if you are not confident in supporting SLAs today, you should plan to design your services and support systems to address both standard and customized SLAs. You will certainly want to understand the parameters of your service network and organizational capabilities before establishing benchmarks for an SLA.

SLA METRICS. Rather obviously, different SLA metrics apply to different types of services. For example, if you are a CLEC operating a broadband voice network, then you will want to include the following metrics in your SLA:

- Call setup time (measured in seconds)
- Service availability (network uptime, 99.9 × percent)
- Network latency, measured by an average of ping tests throughout the month (should be less than 100 ms roundtrip)
- Packet loss, measured over all packets sent during a one-month period for entire customer base (typically around 1 percent or so).
- Billing accuracy (guarantees on delivery, accuracy, and handling of customer disputes)
- Customer service, answer within x minutes

If you are offering a managed voice and data service, your SLAs must include the following major components:

- Service availability (network uptime, 99.9 × percent)
- Security of data while it stays within your network (the customer must know that if data travels across multiple networks, then you have no control over the quality of service from those networks). Security assurances may require you to put in place configuration monitoring mechanisms, access logging, and intrusion detection policies.
- You should be prepared to offer some guarantee on latency (the time necessary for a packet of data to travel across a network; typically less than 100 ms roundtrip). Note that if the customer's data travels across multiple networks, of which yours is only one, then you may be setting yourself up for penalties if you offer an end-to-end SLA or a guarantee of latency or throughput for end-to-end packet transmission. Latency guarantees work well if you own the backbone.
- You must provide necessary call detail to do accurate chargebacks for interdepartmental billing. Most customers will want to know that you can support that level of detail in a timely manner.

· Packet loss, measured over all packets sent during a one-month period for entire customer base. Like latency, packet loss is a relevant metric if you own the backbone.

· Billing accuracy (guarantees on delivery, accuracy, and handling of customer disputes)

· Customer service, answer within x minutes

Table 5.5 shows an example SLA for the data portion of your offering. It can be tailored to your specific needs.

TABLE 5.5   Example SLA for the Data Portion of Your Offering

SERVICE LEVEL AGREEMENT	RESPONSE LEVEL	POSSIBLE REMEDIES
Network Change Request	24 hours	Minor credit
User Change Request	24 hours	Minor credit
Monthly uptime	99.99% uptime (8 minutes of scheduled downtime per month) [Less than 2 hours scheduled downtime per year]	Larger credit
Minor network attack	Notification via local console alert	Larger credit
Major network attack	Customer contact notification via e-mail or paging	Larger credit
Network breach	Resort to predefined escalation procedure to resolve problem	Major credit
CPE hardware failure	1 or 2 business day replacement with configuration	Major credit
Customer problem resolution	Call back within 4 hours, trouble ticket issued same business day, procedure triggered to resolve problem same business day	Larger credit

Part of the SLA must outline what compensation customers will receive when these benchmarks are not met. This is frequently defined in terms of service credits as a percentage of the monthly charges. For example, MCI gives a one-day credit (minor credit) for any ten-minute outage, or a three-day service credit (larger credit) for a three-hour outage. Sprint provides a three-day service credit to small businesses if its dedicated Internet access availability dips below 99.99 percent.

Williams Communications, a broadband provider for bandwidth-centric customers, offers an enhanced SLA for its Dedicated Internet Access service through its high-speed OC-48 Internet protocol network. The SLA guarantees average monthly roundtrip latency of 55 milliseconds and provides for 99.999 percent network availability, with a packet loss of less than 1 percent. The company's SLA is backed by monetary guarantees that grant customers days of credit should performance levels not be met. Customers will essentially receive daily credit for applicable locations should latency exceed 55 milliseconds average or 100 milliseconds maximum roundtrip, network availability fall below 99.999 percent after an initial grace period, or packet loss exceed 1 percent.

You will want to define exclusions to your SLA clearly. Common exclusions include network outages that are direct results of natural disasters (often defined as acts of God such as earthquakes, floods, tornadoes, etc.). Other outages that are not under your direct control or sphere of influence (for example, the ISP network or PSTN if you are using partners for those components) should be excluded as well.

CUSTOMER PREMISES APPLIANCE/CPE DEVICES

If you are including a CPE or an appliance in your service, you may want to outline key configurations or features of the CPE. For example, you are offering a suite of managed services, which you would manage remotely from your network operating center (NOC) or data center. All these services could be delivered on a single, integrated piece of CPE, such as a multiservice appliance. Alternatively, several different CPE may be used to deliver managed services, such as:

- Firewall/VPN appliance
- IAD with built in NAT firewall and VPN
- E-mail server appliance

If you aren't bundling the CPE, you must outline what the requirements are for customers to provide their own. You will probably want to include options between two or three different CPE models as part of the service, to keep things simple for customers.

The CPE section should also outline the requirements and process for installation. Specifically, is the customer responsible for setting up and configuring the CPE appliance? What about on-site wiring? Any other considerations should be discussed in this section.

ORDER PROCESSING AND PROVISIONING

This section of the service definition should describe the process from customer sign-up through installation and service activation. If you are supporting the service with a project manager to coordinate scheduling and the service provisioning across your organization, then describe the steps and process here.

You could also, or alternatively include, a description of the ordering process. This is especially helpful for organizations that only see one piece of the puzzle.

SERVICE PRICING

Many providers take a "me-too, cost-less" approach to their pricing by offering a service package identical to that of the largest competitor in the area but at a 20 or 30 percent discount. Still others lead with "free service" in an effort to capture customers early with the hope of upselling other services. Urban Media's *pièce de résistance* was free high-speed Internet access to small and medium business customers, but the company got stuck on how to sell those customers other services to recoup the costs of the free bandwidth. The upsell revenues never materialized and the company was not able to cover its customer acquisition costs, let alone support any other operating costs. The lesson here is: If you choose to lead with a "free service" you should 1) have an alternate revenue source, and 2) examine your alternate revenue source closely to ensure that it covers not only your operations costs but also your customer acquisition and support costs. Or you may, like Urban Media and many others, regret it later on. The same principles apply to heavily discounted pricing packages such as "free for 90 days," and so on.

Your pricing structure and rates can, and should, be strategic tools to help you achieve your service goals. You can be more successful at attracting your "ideal" customers by thinking about their service usage profiles and tailoring your pricing to encourage those behaviors. Above all, keep pricing simple and easy to understand. Pricing models are increasingly becoming more flexible, and you have several choices available for your pricing schemes.

PRICING OBJECTIVES

Your pricing strategy is a key component of your business plan, and therefore must align with your overall business goals. Be careful that your pricing does not undermine your company's key corporate or service objectives. For example, are you positioned to command a premium price or do you want to be known as a commodity vendor competing on price? Be careful of the latter, because you may have a hard time later expanding into higher-end, pricier services or selling service and quality to your customers. Here are five key questions that will help shape your pricing strategy:

1. *What role does pricing play in your market share strategy?* You can use price to be known as a quality supplier, choosing to target premium customers only, perhaps with a high-priced offering that includes a premium SLA; or use a low-priced service to gain rapid widespread adoption among all customers.

2. *What does your customer profile look like?* How do customers weigh pricing relative to other factors such as brand, reliability of service, and new features? Do they value new features and innovation, or price? Conversely, what image are you projecting or planning to project to existing and potential customers? Answer the question "Why do customers buy from us?" and be sure that your service pricing supports that response.

3. *What competitive pressures exist in your market?* What are competitors charging for similar services, and what does their pricing structure look like? How do your proposed options measure against these—what are the pros and cons? You will want to analyze the competitive service bundles.

4. *What impact will your pricing scheme have on your billing and OSS system?* If you are considering a different or innovative billing model that requires extensions to existing billing systems, how much will you have to invest in the new billing infrastructure? Usually the biggest gating factor of a new billing model is the limitation of existing billing systems and back office support.

5. *Do you plan to offer service bundles?* If so, is there a discounted, promotional, or free component in the bundle? Is it voice or data? How does it impact your revenue and profitability projections? You must factor the discounted or free service or other price promotions into your pricing strategy. If you give away the barn, ultimately customers will suffer as well because of service interruptions related to your troubles.

SETTING SERVICE PRICING

The actual process of assigning rates to your service portfolio is fairly straightforward. The process can be divided into two areas:

- Determining the appropriate billing model for your service
- Setting billing rates for your service

It sometimes helps to assign sample rates to a billing model to evaluate its pros and cons. This can also help show how a proposed combination is meeting your corporate and service goals. You will probably want to examine several different options to see which fits best. This iterative process takes time in the development period, but it is well worth it.

DETERMINING THE APPROPRIATE BILLING MODEL

Your billing model is based on your service components. There are two types of service components: recurring, and nonrecurring or up-front one-time charges. For the service components for the managed voice and data example described in the previous section, the up-front one-time charges would be applied to the following:

- On-site installation of equipment
- The customer premise device (CPE appliance with integrated security features) installation and setup
- Voice service activation

The recurring service components for this example are:

- Six voice lines and associated usage (local, long distance, international)
- Calling features: conference calling, caller ID, voice mail
- High-speed data connection for Internet access
- Basic remote Web hosting
- E-mail management service
- Remote access secure VPN and branch-to-branch secure VPN service
- Basic or premium SLA

In the area of application hosting, many providers, such as data center and Web hosting service providers, are creating higher-value hosting services by offering a full set of network integration and management services. Service elements of a managed application hosting service could include:

- Server and operating system selection, configuration, and ongoing management
- Back-end network integration
- Complete system and applications monitoring

These managed application hosting service offerings may be provided to customers who have sophisticated Web or e-commerce requirements but do not have either the IT expertise or adequate IT resources to cover the management functions.

BILLING MODELS: FLAT RATE, USAGE BASED, VALUE BASED, AND HYBRID

In the past, users have been conditioned to a few fairly standard billing methods from carriers. Billing models, however, seem to have come full circle in telecom. Service providers started out with usage-based pricing, which is pricing based on the number of minutes used. Voice customers got accustomed to usage-based pricing, then the advent of multiple services and the lack of billing systems to measure accurately and bill for these services started the trend for a flat-rate billing model. The evidence is that carriers may be switching back to usage-based (although more sophisticated) billing models again. Variations to the usage-based models, such as billing by packet or type of traffic, may be used.

Overall however, flat-rate pricing is still the most common way of billing for data services and is widely used for local and long distance voice service as well. The easiest to implement, flat-rate billing applies a fixed fee to a service, regardless of the amount of usage. You may have different levels of service, such as a basic service and an enhanced service, each with a different monthly fee. Flat-rate pricing is attractive because it is predictable and easy for customers to understand. Flat-rate pricing is also easy to implement from a back-office support systems perspective as well, since measurement and billing are simplified.

In the case of voice services, flat-rate pricing usually moves to competitive usage-based pricing at a preset crossover point. For example, you can offer 1,000 minutes of long distance for a set fee, and a preset per-minute charge for all minutes after that.

Interestingly, afraid that they may have missed out on incremental revenue opportunities with flat rates, some service providers are moving back to usage-based billing, or, more commonly, hybrid structures using a mix of flat-rate and usage-based pricing.

The hybrid structure that most long distance service providers use consists of a mixture of flat-rate and usage-based charges. A good example is international calling plans offered by Sprint, MCIWorldcom and AT&T: users pay $xx a month (usually $3 or $5 per month) to be on a special plan; they then are charged discounted per-minute rates depending on the call destination. This is a common structure for voice services.

As data services become more advanced, billing models will become both more fluid and more flexible, driven in part by the desire to derive incremental revenue from perceived value, and in part to keep pricing simple. With so many choices in billing models on the horizon, service providers will be able to use pricing as a differentiator and competitive weapon. They will, however, have to invest in next-generation billing systems that measure the various types and characteristics of service usage, focusing on metrics such as packets or type of traffic.

As an example of a new type of billing model, value-based or event-based pricing is only just starting to be used by data service providers to charge for premium or "pay-per-use" services such as streaming video events, distance learning or training, network resource utilization, and the like. Other billing models include tiered pricing such as for burstable bandwidth; and additional levels of services such as gold or platinum.

Keep in mind that the type of billing model you use will influence your traffic loads during peak and off-peak hours and can impact your network utilization and costs. For example, a completely flat-rate structure tends to favor and attract heavy users, because there is no usage penalty if rates are set reasonably. If you don't have a corresponding package for off-peak customers, your network utilization may look unbalanced over a 24-hour period or across a seven-day period.

SETTING BILLING RATES FOR YOUR SERVICE

Before setting service rates, you should conduct some research, on competitive pricing in your market space and on your costs to support the service. The financial analysis you did when initially building your market entry and service strategy (Chapter 4) should help a great deal here in terms of identifying fixed and variable costs for deploying a service.

One recommended approach, such as that described by TeleChoice, is to use the following exercise for setting service rates:

· *Competitive analysis.* What are the charges for competitive services, and what are their billing models? How do your proposed options measure against these—what are the pros and cons?

Table 5.6 shows what your competitive pricing scenario may look like if you are offering broadband voice.

TABLE 5.6 Sample Competitive Pricing Scenarios

SERVICE PROVIDER	PRICING	PRODUCT SET
CLEC-A	$400–$650 per month	6 voice lines 1,000 long distance minutes 50 email addresses Web hosting 1.5 Mbps DSL
CLEC-B	Flat rate: $649 per month	8 voice lines with CLASS features 6,000 anywhere/anytime voice minutes 768 Kbps SDSL circuit
CLEC-C	15%–50% below ILEC service rates	All ILEC products

- *Substitute analysis.* What are the charges for other voice services, such as POTS, VoIP, Centrex, or those services that your target customer is using today?
- *Cost-based analysis.* What are the fully loaded costs to support the service, including operations expenses, capital costs, customer acquisition, and sales and marketing costs? Do your proposed revenues per customer increase appropriately?
- *Portfolio analysis.* How does the pricing for your service line up against other voice services in your portfolio? Do they encourage customers to choose the appropriate service?

If you are leading your offer with a combined voice and data service package, then you'll need to look at each of these with respect to all of the discrete service components. You should analyze competitive service bundles as well.

Once you've gathered the data for these four different analyses, build a spreadsheet to compare each of these against each of your proposed billing models and rates. Incorporating your expected customer adoption will help you to determine which approach fits best.

ASP BILLING MODELS

As with planning a market entry strategy, if you are a hosting service provider or ASP, your pricing model is likely to be different from that of a CLEC's managed service pricing model, and it may require different considerations. ASP services (at least for large enterprises) are integration intensive, often requiring customization; pricing can be based on the following service areas:

- The application itself (including licensing fees)
- The level of customization and integration to remotely host the application
- Annual maintenance fees

These service areas can be very different for e-businesses, and for small, medium, and large enterprises. For instance, large enterprises have more choices available in hosting applications. Because there is usually a well-staffed internal IT organization and IT budget, the overriding factor in utilizing an ASP is wanting to get rid of noncore tasks, such as maintaining hardware and the day-to-day administration of running applications that may distract the IT organization from more critical core activities. The enterprise may want to buy and retain control of the application and associated licensing, and outsource the hosting and maintenance portion of the application. Large enterprises also frequently want customization.

Small and medium enterprises, on the other hand, have limited IT staff or expertise and are more likely to rent the entire package (application + hosting + maintenance) for a fixed price per month from the ASP. Small businesses typically do not want customization, usually to keep costs low, and because the types of applications this segment tends to use (payroll, e-mail, calendaring, etc.) usually do not benefit much from customization. From a small-business ASP's perspective, getting into application customization can be resource intensive and may not provide good returns. In all but the large-enterprise segment, volume sales of fixed-fee rentals are the key to generating profits in the ASP business.

Various pricing models can be applied to application hosting. You can use a tiered pricing structure to price on a per-user basis (the most common pricing model), or price on a per-transaction basis or per-usage basis. The latter two are best suited for event-driven "pay-per-use" applications such as videoconferencing, bandwidth on demand, storage on demand, entertainment, financial transactions, among others. Clearly, larger enterprises offer greater incremental revenue opportunities for ASPs through these value-added services. When targeting large enterprises, for example, you can charge for additional services such as bandwidth and storage space made available to the customer. You can also derive revenue from bundling related services, such as an e-commerce package for Web hosting customers. Colocation service providers, for example, typically apply additional charges for rack space leased by customers. Similarly, advanced hosting providers are setting higher fees for hardware integration and complete system management. Customized hosting (required by large enterprises) requires integration with legacy systems or highly customized designs and configurations; consequently, such hosting services can command the highest premiums, to the tune of $250K per Web site per month. A number of ASPs serve the large-enterprise market: EDS and SAP are such ASPs. The most common applications supplied to this segment are ERP and CRM packages from software vendors such as Siebel, Oracle, and others, along with a smaller number of customized Web and e-commerce hosting applications supplied by companies such as Broadvision.

While large enterprises will easily stomach such premiums, small and medium businesses with average IT budgets of about $25K to $100K per year for all their IT spending, require a different level of service and pricing tailored to their needs and budgets. Basic Web hosting typically runs around $45 per Web site per month for small businesses, with premium Internet content, e-mail, and e-commerce as an additional charge. The ideal pricing model for a small business is a fixed monthly fee for the complete ASP package, based on the number of concurrent users (not named users) of that application. For medium businesses (100 to 500 users), a tiered pricing structure based on user buckets of 50 or more concurrent users can work pretty well. In addition, the ASP typically charges an up-front set-up fee for installation and integration. Depending on the type of services you offer and the buying constraints of your business customers, you may want to consider spreading the set-up fee (which can be $5,000 or more) into 12 monthly payments, or simple bundle it into your monthly fee by raising the monthly price over the term of the contract. This can put constraints on your own cash flow, however, and should only be done if your business model can survive the delay in revenue recognition. ASPs who have a significant amount of cash on hand can benefit from this type of pricing model, because it lowers the cost barriers for small businesses and can accelerate adoption.

In a 2000 survey by Current Analysis, users indicated a preference for alternative pricing models, such as a flat monthly fee for an entire company, or per-user fees assessed on number of concurrent connections, rather than named users. The pricing models of ASPs are still in a fluid state, because both the application delivery mechanisms and business models of ASPs are changing to adapt to the economic times and the inevitable consolidation that is occurring among different players. Current Analysis summarized the pricing model currently used by most hosting providers in Table 5.7.

TABLE 5.7   ASP Pricing Model (*Source: Current Analysis, 2000*)

PRICING MODEL	PRICING ELEMENTS	TARGET CUSTOMER
Up-front charges plus monthly fee	Up-front fees for integration, consulting, and/or customization— option of transferring software and hardware assets to customer after a defined period	Larger companies with more complex implementation needs
	Monthly fees based on usage, number of named users, or number of concurrent users	
Bundled monthly charges	Up-front cost included in monthly payments	Smaller companies with minor integration and installation needs or companies interested in preserving start-up capital
Flexible/revenue sharing	Up-front cost but no fixed monthly—ultimate pricing driven by the total success of the site (including number of users and transaction volume)	Companies running e-commerce or business-to-business commerce sites
	Minor or no up-front cost—total pricing based on the transaction volume and ASP service performance	

PARTNERSHIPS— EFFECTIVELY WORKING THE VALUE CHAIN

Your service strategy calls for delivering a rich portfolio of services. It is very likely that you have the core competency and resources to deliver a subset of those services. Wholesaling or partnering is a quick way to fill the gaps in your portfolio and provide the extra edge that will distinguish you from your competition. In today's environment, you will also want to consider partnering to get access to critical network resources such as network integration, management, and planning. This chapter discusses the importance of partnerships in telecom, and how to go about identifying and forming appropriate partnerships in the service delivery value chain.

Nowhere are partnerships more critical than in telecommunications networks, where every piece of the network is owned or operated by a different entity. Partnerships are the conduit to access those pieces. Some companies are formed around a partnership strategy from the start, whereas others only require partnerships as they expand beyond their initial business. Whichever camp you fall into, this chapter provides insights to help you recognize and achieve your partnership goals.

MARKETING PARTNERING VERSUS STRATEGIC PARTNERING

The type of partnerships that you need to cultivate are not the "marketing" variety, which have sprung up all over the high-tech world and have maligned the word "partnership" to some extent. Typically rather superficial in nature, marketing alliances arise out of a need for companies to mutually create marketing hype and inflate their respective products' functionality and value. These relationships exist mostly in boardrooms and marketing literature; other than loosely defined "alliances," there's not much to it—the products from the partnering companies are not integrated with each other, nor is tight product integration a requirement for the partnership to be created.

In telecom, partnerships solve a very different problem and therefore play a much more critical role. Telecom partnerships tend to be more strategic in nature, almost always put in place to achieve very specific network or service goals, rather than marketing goals. The fact is, your partners are there to fill a very real need in your network, whether to deliver a value-added service or product to your customers, to manage your data center for you, or to perform other critical interdependency functions, such as backbone transport setup or network management of your access network. Whether through an SLA, a consolidated service bill, a call center, or a service-ordering portal in a building basement, your partners touch your customers and your reputation. Unlike a hands-off marketing alliance, telecom partnerships call for complete trust and a tight working relationship.

PARTNERING AS A STRATEGIC ASSET

Aside from filling a gap in your portfolio, what are the benefits of partnering? When does it make sense for you to partner? With whom should you partner? How should you structure your partnerships? What role will these partnerships play in the end-customer's purchase decision?

The answers to these questions may lie in your value chain definition—the type of services that you plan to deliver to your customers and the entities you will need to interact with to make that happen. Other than building it yourself, there are two ways to get quick access to networks and services: acquisition and partnering.

ACQUISITION AS AN ALTERNATIVE TO PARTNERING

Acquisitions abounded in the late 1990s, and when the markets slowed, several service providers felt these acquisitions drag down their balance sheets. For service providers with clear-cut business strategies, rock-solid financial models, and a good head for bargain hunting, acquisitions can be a way to beef up your company resources and products overnight.

Acquisitions can also be profitable if executed purposefully and in a timely manner. One company that has rapidly, and apparently successfully, acquired its way into the Web hosting business is Allegiance Telecom. Once Allegiance decided to ramp up its Web hosting business, a flurry of acquisitions followed. Within a period of eight months the company had completed its mission of developing a full-fledged Web hosting business.

The trail of acquisitions goes like this: in October 2000, Allegiance acquired Virtualis Systems, a Web hosting company and applications service provider. This first acquisition allowed Allegiance to expand its service offerings to include Web hosting and Web site management capabilities. In early 2001, Allegiance acquired two regional ISPs: InterAccess in Chicago and CONNECTnet in San Diego. Both acquisitions provided Allegiance with an established customer base and the opportunity to cross-sell an integrated package of communications services. On April 16, 2001, Allegiance announced its acquisition of HarvardNet, a Web hosting company targeting small business customers in the eastern part of the United States. Then Allegiance announced that it had acquired Adgrafix, a Web hosting services company headquartered in Sudbury, Massachusetts. These new acquisitions solidified the company's East Coast footprint. Finally, Allegiance consolidated all the acquired companies into a single centralized Web hosting division, and on May 2, 2001, unveiled Hosting.com as the brand name for this division, which now operates as a wholly owned subsidiary of Allegiance Telecom. Hosting.com offers a complete line of Web hosting and Internet connectivity services targeting the small- and medium-sized enterprise (SME) market.

Clearly, Allegiance executed well on its acquisition spree, but I caution young emerging start-ups on pursuing this route. Companies are always to be had on the cheap, and obvious benefits can be gained from adding a ready-made team and product suite to your asset base literally overnight, particularly if the purchase terms are attractive and the technology or service of the acquired company is sound. But this is a road fraught with perils, as many a telecom company has discovered the last couple of years; particularly for a small company, the cons far outweigh the benefits. For one thing, too many companies end up overpaying for assets. Second, postmerger integration of the two companies' products and people can be a complex and lengthy task and can easily distract the new company, thus resulting in lost sales and market opportunities. Third, if you are acquiring a company to get into a business you know little about, chances are high that you will botch it. Last, all things considered, the odds are 50-50 that you will do this successfully. Just as many mergers have failed as have been successful. For most small start-up telcos, partnering is the better alternative to rounding out your line of business and service portfolio.

PARTNERING

Two types of partnerships are needed for delivering telecom services:

- Partnering to get access to complementary services
- Partnering to get access to network resources

PARTNERING FOR COMPLEMENTARY SERVICES. Why would you want to partner for value-added services? The answer is obvious: you obtain access to value-added applications and services from other companies, in a relatively short period of time, that would have taken you much longer to develop and market on your own. You are able to pass on these complementary services to your customers in a timely manner and generate additional revenues. Expanding your service portfolio offers several benefits to you:

- Additional revenue streams
- Differentiated service offerings
- Opening up new markets
- Promoting customer stickiness and retention

You can use a number of partnering strategies to reach your customers. Both reselling and private-labeling complementary services hold several advantages. Reselling a service does not burden you with technical support issues, and you are typically compensated in the form of a flat sales commission, which can be as much as 10 to 30 percent depending on your involvement. There is also minimal integration involved in reselling. Private labeling makes you the first point of contact for the customer and responsible for resolving technical issues. While private labeling can involve more work than reselling, it is preferable because it allows you to retain customer contact for the complementary service. Private labeling also gives you a measure of control over pricing and feature enhancements to the complementary service, thus providing pricing and distribution flexibility.

In both cases, however, you gain access to several time-to-market advantages:

- You can respond quickly to changing market needs
- Your start-up costs are lower
- There is no need to hire additional support staff (although private labeling could require retraining some of your support staff)

· You can offer a wider variety of services and generate incremental revenue
· You have an immediate competitive advantage

The more difficult question is, which services in your portfolio are best delivered through partnering? A good place to start is· to understand the time-to-market, competitive, and financial pressures of your service launch. For example, some services that you may be investigating for your portfolio may include voice, Internet access, VPN, Web hosting, and ASP. Out of these, the most likely candidates for partnering are voice, Web hosting, and ASP, because these services require substantial investment in infrastructure, training, and staff expertise in policy planning, network management, and operations support. Putting the pieces in place to support this may take 12 to 18 months, time which you do not have.

One of the most challenging to implement, ASP services are best delivered through partnering with an ASP or ASP aggregator. There are various ways this can be accomplished, and these are described below.

PARTNERING FOR ASP SERVICES.   The struggling ASP industry is set for revival this year and offers great opportunities for carriers to participate in its growth. With the economy in the doldrums, renting applications from ASPs offers opportunities to the small and medium business for lowering costs while continuing to reap the benefits of access to necessary business process applications such as HR, payroll, accounting, and CRM. If you are a CLEC or a BLEC, wholesaling ASP applications offers you a strategic way into application hosting, which is the next logical step to becoming a "Home Depot of Telephony."

Traditionally, ASPs have targeted the large-business community with business process applications that require substantial integration and customization. Now, however, they are making these available to small and medium businesses in a prepackaged form. In the large-enterprise market, the ASP owns the customer relationship, because it serves these businesses directly. In the small- and medium-business market-space, however, the smaller ASPs do not have a direct customer relationship and are therefore turning more and more to retail service providers—usually those BLECs and CLECs who have primary contact with the customer. Since the application requirements of the small and medium business are less complex, some carriers in this space are becoming ASPs themselves by bypassing the ASP and instead partnering directly with the independent software vendor (ISV) to deliver hosted applications to customers. British Telecom recently announced its BT Ignite ASP division, which offers Oracle's e-business products as an ASP service. In this partnering arrangement, Oracle provides the software and support, whereas BT provides the hosting, hardware maintenance, and service delivery to customers.

To partner with an ASP, you must first understand the ASP's business practices and value chain—at least for those ASPs who are more than application developers and have developed a business plan for deployment.

THE ASP BUSINESS MODEL.   The ASP hosting business is complex in nature. The model is essentially that of a managed service, with a primary focus on net-hosted business process applications (such as HR, payroll, CRM) that the ASP owns or buys from ISVs such as Oracle, Vantive, or Siebel. (Some ASPs now offer productivity applications for small businesses, such as messaging and collaboration, invoicing, document management, and so on.) ASPs are also going after "pay-per-use" applications such as audio and videoconferencing, storage and backup, and other event-based applications that can be billed on a per-usage basis. These applications reside in the network, in a centralized location such as a data center, from which business customers can access them over a private network or a secure VPN link over the Internet. The user interface at the customer premise is typically an Internet Web browser with an application icon or an application window. Figure 6.1 shows the comprehensive nature of the ASP business model.

The ASP offers contracted services to end-users either on demand or on a monthly subscription basis. It is responsible for

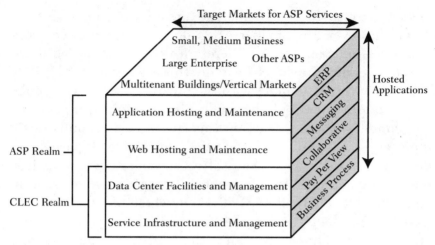

FIGURE 6.1   The various facets of the ASP business, and where the ASP may desire CLEC services.

integrating related software, hardware, and networking technologies to deliver the application service in lieu of the customer's owning and managing these applications. For small businesses, these services are typically prepackaged with little or no customization.

THE ASP VALUE CHAIN.   The ASP network generally involves various players for service delivery. These can include ISPs, network access providers (this role is usually filled by the CLECs), competitive access providers (CAPs), systems integrators, backbone providers such as Qwest, and long distance carriers (IXCs). An ASP thus has to deal with various partners to host applications successfully. The ASP may or may not own or manage its own data center, but whether it outsources service infrastructure or not, the ASP generally will deploy, manage, and enhance hosted applications. Figure 6.2 is a sample ASP network, showing the various service providers and the relationships between them.

HOW YOU CAN PARTICIPATE IN THE ASP BUSINESS.   There are over 2,000 ASPs today, but there are two forces working against this number. First, most of the smaller ASPs do not have enduring business models; and second, the market cannot sus-

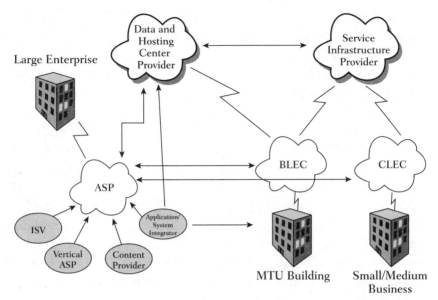

FIGURE 6.2   The ASP network.

tain that many players and will force the weaker ones out of business. As a result, the ASP industry will undergo some serious consolidation and liquidation in the near future. Nonetheless, business process and productivity applications abound, and there is growing demand for subscription-based business applications in the small- and medium-business community. The market for ASP services is poised for tremendous growth over the next few years.

There are many ways you can get into the ASP business, ranging from reselling applications to providing comprehensive network resources to ASPs. Here are the different roles you can play:

· *ASP reseller.* The reseller of packaged ASP applications is the simplest role. In this case you are not private-labeling the service, but simply offering your customers an independent ASP solution for a piece of the revenue, typically less than 20 percent. You can boost revenues by providing your own value-added services on top, such as desktop support or application integration. Although revenues from reselling are limited, this is a quick way to expand your portfolio. The simplicity of this model is appealing, and some service providers, (even incumbents such as BellSouth) have chosen to go this route. Almost any ASP application can be resold, but consider those that are horizontal and offer larger potential for revenues, namely productivity and some business process applications.

The downside to this approach is that you may find yourself reselling multiple ASP offerings, because a single ASP is not likely to fill all your customers' business needs. Assuming your customers want to deal with multiple ASPs, you could still be creating an integration and support nightmare for them, and this could complicate and adversely affect your relationship with your customers. A potential solution is to partner with a large and established ASP, such as Corio, which offers multiple preintegrated applications and functions almost like an aggregator.

· *Vertical ASP aggregator.* This approach allows you to integrate applications from different ASPs, and you can create vertical packages for different segments. A vertical approach to aggregating applications is better than a horizontal one, because customers more often than not want to buy applications that tie into their business processes. These tend to be vertical in nature, differing by segment. Horizontal applications, such as e-mail, have broad appeal across multiple segments but come with a lower price tag and increased competition. Adding the vertical component allows you to differentiate yourself and also charge a premium for the unique touch you are providing. The applications should be accessible online to customers through a Web portal with a single logon ID. You have now become a one-stop-shop solution for segment-specific ASP applications, providing a single point of contact to customers for integration, billing, and support.

This approach is more complex and resource intensive but more appealing to customers, because it minimizes integration and support issues. Most customers do not want to deal with multiple ASP contracts and multiple support contacts. Most customers also want an ASP that understands and caters to their business processes.

Web-based productivity and business process applications are ideally suited to online aggregation. You can negotiate favorable wholesale rates from individual ASPs and still charge a premium for delivering a complete integrated solution. The challenge here is that you will have to research hundreds of small ASPs, and negotiate appropriate contractual agreements that protect you and your customer. The other danger is that several of the small ASPs are likely to fail in the coming year, so you will want to examine company fundamentals closely before getting into a business relationship.

Verticalizing ASP applications requires a balance between standardization and customization. A rule of thumb is that 80 percent of your application basket should be standardized (i.e., applicable horizontally—such as e-mail and calendering) and acceptable to the majority of your customers. Twenty percent could consist of applications specific to a vertical segment.

If you like the appeal of ASP aggregation but don't have the resources to integrate and deliver applications, you could consider wholesaling from ASP aggregators and private-labeling the brand, billing, and support in your own name. The drawback to wholesaling from an aggregator is that you will likely not be able to offer vertical packages, and may not be able to negotiate favorable pricing.

· *Application infrastructure provider.* Service providers who provide the service infrastructure are called Application infrastructure providers (AIPs). This is by far the most complex ASP scenario of all to implement. Many ASPs outsource the service infrastructure to service providers and focus only on the application implementation and management. This presents an opportunity to service providers to supply the networking infrastructure and server farms where applications and Web sites are hosted.

As an AIP, you can provide varying degrees of infrastructure services to ASPs:

–*Application management service provider.* An application MSP provides the configuration and management of hardware and software platforms for application hosting.

–*Managed data center provider.* This is essentially the role of an infrastructure service provider who builds data centers with the corresponding networking infrastructure and server farms for Web and application hosting. As a managed data center provider, you can offer Web hosting services along with colocation services consisting of cages, racks, and infrastructure management for ASP applications, thus making it attractive for an ASP to rent from you.

–*ASP platform provider.* Wholesaling an end-to-end data center and network to ASPs for application hosting is the key here. In addition to the data center facility, you will need to provide the configuration and management of hardware and software platforms used for application hosting. Finally, becoming an ASP platform wholesaler requires you to bundle a fully managed billing, provisioning, and customer care system.

–*Content delivery management provider.* An ASP is going to require reliable content delivery management of the hosted applications over the shared infrastructure and backbone. Data centers are typically connected to the Internet, ISPs, and other affiliated data centers to create a shared backbone infrastructure for application traffic. Guaranteed bandwidth, security, and 24/7 network support often are requirements in the SLA, particularly for running critical applications such as payroll and CRM. As a content delivery management provider, you can offer ASPs efficient management of the backbone infrastructure, monitoring backbone network performance, availability, and security to ensure application content delivery to customers.

PARTNERING IN MULTITENANT (BLEC) MARKETS

To a BLEC, partnerships play a key role in the business model. There are a variety of partnering activities associated with establishing and operating services to multitenant buildings. For starters, this particular segment requires you to partner with the property owner, often at an executive level, before you can begin marketing and delivering your services to tenants. The property owner is an important entity who holds the keys to the building and can grant you special privileges or access rights—these must be negotiated at the executive level, of course.

To penetrate the MTU market effectively, you must engage in a variety of activities, all of which require a wide range of expertise. It is unlikely that, as a single business entity, you will want to engage in more than one or two of these functions. More important, each activity impacts the bottom line and is crucial to your success.

MTU END-TO-END SERVICE DELIVERY CHAIN

A BLEC must undertake several steps to provision an MTU building with broadband services:

- *Initial building evaluation.* Several players, including the property owner, the BLEC, and wiring contractors or consultants perform building evaluations for physical access to the in-building wiring and riser, to assess demand for services, and provide appropriate rights of entry to relevant parties.
- *Mini-POP in building basement.* The switching equipment or concentrator is installed in the building basement for access to in-building wiring. Wiring may be pre-provisioned to the different floors of the building for the quick turn-up of tenant subscribers.
- *Local access infrastructure.* Most BLECs lease local access facilities from the ILECs. BLECs that own their own local access facilities are usually deploying fiber to the curb and do not have to rely on the local access infrastructure of the incumbents. A significant benefit of owning your own local access facilities is that you have greater control over service provisioning and can roll out services much faster than if you hmust deal with the bureaucracy of the ILECs. Avoiding the ILEC also means eliminating loop qualification issues that slow service deployment. As a BLEC, you can plan to deploy

your own Gigabit Ethernet switching and routing equipment, and lease dark fiber or construct your own local fiber network.

- *Regional switching center.* The BLEC must backhaul traffic from a given metropolitan region into a single switching center. At this point, data traffic is usually aggregated and handed off to an ATM switch or router. BLECs offering facilities-based voice services may also own a Class 5 switch.

- *Backbone infrastructure.* The backbone provider must be selected and the backbone installed. This requires physically connecting the equipment in the basement POP as well as connections in the regional switching center. The BLEC needs to establish a relationship with the backbone carrier to facilitate trouble resolution.

- *Network operating center (NOC).* The BLEC's NOC administers the network including system security, performance monitoring, bandwidth administration, network configuration, network problem detection and resolution, and often billing and cost accounting. The NOC typically houses a network management system, a security system, and sometimes a billing system.

For all of the above, a BLEC must implement not just components of the physical infrastructure, but the less-tangible partner relationships as well. Any provider who tries to do it all alone is going to run into a lot of problems. Therefore, the BLEC must mature its relationships with the CLECs, ILECs, and IXCs who supply it with transport and other wholesale and value-added services. The BLEC is very dependent on this latter pool of companies for its required network infrastructure. The BLEC will also need to establish relationships with ISPs and ASPs for Internet access and value-added services.

The BLEC's service delivery chain is shown in Figure 6.4.

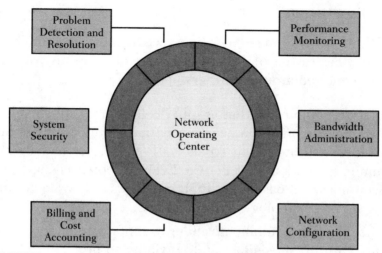

FIGURE 6.3 The various tasks performed by a NOC.

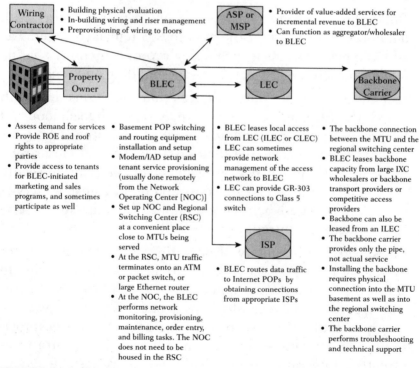

FIGURE 6.4 The roles of various entities in supplying broadband services to MTU buildings. The BLEC owns the primary contract with in-building customers and works with different players in the value chain to service the customer.

BLEC PARTNERING PRACTICES

As a BLEC, you will need to follow several guidelines for delivering services to the MTU market. Most of them, as you can see, are centered around partnerships:

· Look for equipment that is physically compact, collapses multiple voice and data access functions into one, and supports differentiated, and typically IP-based, services over a common network infrastructure. Evaluate potential suppliers based on a set of basement–POP requirements. Avoid getting hung up on technology "wow" factors.

· Ensure that the equipment vendor has interoperability agreements with other vendors and is willing to provide proof of interoperability. The goal is to demonstrate a successful technology or lab trial followed by a limited customer roll-out or full-scale deployment, depending on the market size.

· Build the necessary partnerships and rights of entry with landlords and REITs for the buildings you plan to target. The BLEC typically needs to establish a revenue or equity-sharing model in exchange for access to the building and tenants.

· Establish the necessary services and relationships with ILECs, CLECs, and IXCs that supply you with local loop and backbone access, provisioning services, and other wholesale services such as local and long distance voice. A closer relationship with these service providers will allow you to secure volume discounts on local access and other transport services, and to private-label, more complex, IP value-added services as these are introduced.

· Investigate the feasibility of operating as an ASP aggregator by setting up a Web-based portal that is easily accessible to subscribers for ordering applications. Extend this "self-care" concept to include multiple statistics, billing logs, and account status and also real-time usage and service provisioning and payment. Develop a suitable billing infrastructure that integrates the various services at a single point. Work with ASPs and other value-added service partners to ensure a consistent level of quality and interfaces.

- Develop vertical packages for the top three or four categories of business tenants occupying the building. The most suitable applications would be those that help tenants be more efficient in their operations. Develop a targeted sales program to educate tenants about the benefits of these packages.

- Focus on developing a partnership strategy that adds component services at a layer *above* the infrastructure and transport. Your primary objective should be to utilize value-added service partnerships to improve time to market.

MANAGING ENTITIES IN THE SERVICE-DELIVERY VALUE CHAIN

ENTITY ROLES

Today's telecommunications picture is a complicated, often interdependent web of entities that sometimes compete with each other and at other times, partner. Some of these entities are responsible for the physical elements of the network. These include backbone carriers who provide Internet backbone services, local service providers such as ILECs, ISPs and data center providers who provide infrastructure for colocation, and Web/application hosting. Some of these entities offer hardware, software, and services for the Internet computing infrastructure, within the broad categories described in Table 6.1.

ENTITY SYNERGIES

As a service provider you have developed a service-delivery value chain consisting of a number of discrete entities. To deliver services to your end-customers, you must interact with the entities in your service-delivery value chain. Your partnerships with these entities serve an important purpose: to accelerate the delivery of a high-quality, reliable, and complete service offering to the end-customer over a secure and reliable network. Conversely, the entities in your value chain have a similar goal: to deliver services to *their* customers. Everyone in

TABLE 6.1   Telecom Companies' Roles in the Service Delivery Value Chain

ENTITY	ROLE
Equipment Manufacturer	Provides telecommunications and switching equipment to service providers
OSS Supplier	Provides Operations Support Systems to service providers
Managed Service Provider (MSP)	Provides managed data services from a network management system (NMS) typically located in the regional POP or data center; this role is often filled by a CLEC, BLEC, ILEC, or IXC
Application Service Provider (ASP)	Provides net-hosted applications from a data center
Internet Service Provider (ISP)	Provides network layer services (IP addressing and routing) for Internet access; may also function as an Internet data center provider
Network Service Provider (NSP) or Backbone transport provider	Provides network and backbone infrastructure services
Local Exchange Carrier (CLEC, ILEC, BLEC): competitive, incumbent, and building-centric	Provides local and long distance voice and data services; also offers broadband connections such as DSL, T1, wireless, fiber, etc.
Inter Exchange Carrier (IXC)	Primarily provides long distance phone services; may also be a LEC or MSF
Application Infrastructure Provider (AIP)	Provides hosting facilities for ASPs; this role may be filled by a one-stop-shop CLEC

the value chain thus stands to benefit from partnering, and natural synergies exist between the various entities in the value-chain.

As an example, consider the case of serving a major metro area with high-bandwidth services. Major metropolitan areas have plenty of dark-fiber capacity available, and more is being installed every day. The challenge facing you is to come up with a deployment strategy that is both rapid and cost effective. You have figured out that leasing the necessary optical capacity and providing fiber termination alone is not enough and will not enhance the bottom line. You are thus opting to package the high bandwidth with an array of services in exchange for a recurring monthly fee. In essence, you are operating as a managed services provider for small and medium businesses.

In this scenario, your customers are typically physically connected via the local network to a regional center or POP, which connects to the Internet through a network service provider's (NSP's) backbone network. The NSP provides the connection to an ISP, which provides both administered IP addresses and the necessary routes to the Internet. Your network operations center (NOC), which is the management component of your service offering, can be physically located anywhere, but usually resides in a regional data center, where you have obtained colocation space. (Some facilities-based CLECs, such as McLeodUSA, are "super" CLECs in that they act as CLEC, ISP, and network service provider.)

In the example above, you partner and cooperate with the fiber backbone providers, switching and storage vendors, network operators, and others, all of whom are a part of your delivery value chain. You also serve as a distribution channel for vendor equipment. You buy the equipment from the telecom equipment vendor, and typically provide the customer premise equipment (CPE) as part of your managed service. The backbone transport providers, switch, and storage vendors are motivated to collaborate with you on initial customer successes in the hope that this will demonstrate the success of fiber in metro areas and accelerate the growth of the optical-access market, thereby maximizing the use of existing fiber, accelerating installation of new fiber, and lowering the price barriers for fiber access. Whatever the individual goals of the participating players, tremendous synergy exists among them.

As another example, natural synergy exists between access and backbone transport providers and data center providers. Because backbone bandwidth is a large cost component for a data center provider, integration with a backbone provider brings economic benefits. Data center providers benefit from reduced transport costs as a result of a strong partnership with the backbone provider, but backbone providers also benefit by being able to "fill the pipes," or secure IP traffic. There are benefits for the network access provider, too, because data center and hosting services will require faster speed and higher bandwidth services at the customer premise. In this case, the synergy that exists among providers can facilitate the bundling of data access, transport, and application hosting services.

These various entities provide one or more of the value creation activities that together will create your value-added service offering. To the end-customer, however, you are the single point of contact—the customer is buying the whole service from you. Your customers do not care about the nature of your partnerships, only about the products and services that you sell them as a result of the partnership. In the absence of owning your own backbone network and infrastructure, forging partnerships at the critical points in the network where interdependencies occur ensures that your customers receive secure and reliable services.

Every service provider's goal is to maximize availability and minimize failures in the network. This can be a daunting task for smaller CLECs, given the number of nodes that traffic must pass through in an end-to-end exchange. This lack of network ownership on the part of smaller CLECs is often the target of larger carriers with a private backbone. In recent ads, for example, MCI Worldcom stresses its competitive advantage over smaller brethren: unlike the smaller CLECs, MCI Worldcom need not transport data and voice traffic over several separate networks, because it owns its backbone network. MCI Worldcom doesn't "hand off" customers' valuable data to other providers, and therefore its network is not prone to service failures. MCI's message is that a single far-reaching network is better than multiple networks strung together, and that the CLEC's services are inherently unreliable because data is handed off across several networks.

CLECs who do not own their own private backbones are at a disadvantage in the short term. Only short-term though, because "hand-offs" between multiple networks are a necessity for future telecommunications networks, and thus these hand-offs *must* become more reliable. As the number of new and specialized services increases, packet voice and data traffic will increasingly travel over multiple networks, frequently owned by different service provider entities. Interdependencies have largely become a way of life in the modern network. How carriers deal with each other's networks influences to a great deal the quality and reliability of service to end-customers. In the absence of defined QoS parameters in the public IP network, service providers young and old are achieving reliability and quality of service by negotiating IP and network management tasks with other partners in the network.

ENTITY PARTNERING PRACTICES

As a service provider, you must develop the means to evaluate potential partners, try out their products and services, and create contractual arrangements whereby you can supply third-party products and services to your customers. You must evaluate the viability and credibility of your partners, particularly when dealing with the smaller equipment manufacturers, ISPs, and ASPs. Some recommended guidelines for working with telecom equipment vendors are discussed in the next sections.

WORKING WITH EQUIPMENT VENDORS

NETWORK EQUIPMENT.   The following criteria should guide your selection of a network equipment vendor:

- Vendor reputation (How many years has the company been around? Can it furnish references? Does it have the ability to support a network of your size?)
- Is the product network ready? Has it been deployed and field-tested in production environments?
- Has the product been tested for interoperability? Most vendors today are establishing interoperability labs to test and demonstrate interoperability between their products and

other components in the network. Ask for proof of interoperability through a live demo simulating network conditions, and certifications where necessary.

· Is the product future proof? Is it based on standard protocols and will it accommodate growth in your network without requiring a forklift upgrade?

WORKING WITH OSS VENDORS

By far the most critical of all systems in your network are Operations Support Systems (OSS). Most small CLECs do not have the expertise to specify OSS requirements, let alone select the right vendor to implement the appropriate systems. An OSS system is the backbone of your operations, and therefore the process for this vendor/equipment selection is more complex and requires substantially more in-depth planning than does other network equipment. Usually a systems integrator is involved.

OSS, in particular the provisioning and billing systems for integrated voice and data services, requires the integration of separate islands of information within the service provider's facilities because a single access facility, such as DSL or fiber, is used to provide multiple services. Legacy systems for voice and data were not designed for future integration, so such a task can frequently be complex, expensive, and time consuming, particularly for a facilities-based CLEC that has to integrate separate networks. If you are planning to deploy packet voice, you will need to overhaul your existing billing and OSS systems to track and manage the billing details for your integrated services.

It is important not to underestimate the importance of this task, because this integration can often be a limiting factor for the deployment of any telecom service, either in terms of time to market or in terms of service capabilities and functions. Depending on your resources (both financial and personnel) and how ambitious your deployment plans are, this task can often take up to a year to perform and can cost millions of dollars.

I interviewed several OSS agencies and systems integrators to try and pin down a workable standard process for selecting an OSS system, which is probably one of the most difficult tasks a carrier faces. I came to the conclusion that determining the exact details of how an OSS system should be selected rests with the experts, namely an OSS integration company or systems integrator such as Pricewaterhouse Coopers. This is not a task that start-up carriers can undertake themselves, unless they have access to considerable OSS expertise in their staffs.

With that in mind, every startup CLEC and service provider must understand what's involved in identifying OSS requirements and selecting the appropriate OSS vendor. The following section describes suggested guidelines for accomplishing this task. These guidelines are drawn both from my own experiences and research, and from my discussions with Paul Roemer (among others), who runs his own OSS integration consulting company, Spectralliance, LLC, for telecommunications providers and the applications vendors who support those firms. Note that the process assumes that there is currently no OSS in house, and that the service provider is in the early stages of starting operations. Established carriers typically bring in a systems integrator to tie existing billing and OSS systems into their new billing modules and ensure that the proposed billing system works with applications in the new network.

Planning the OSS Vendor Selection Process. Most executives know they need a back-office support system; the problem is that their knowledge of which one to buy and what they need this system to do is very limited. They know they need to be able to generate a bill and to be able to respond to customer queries, but that's about it. The following guidelines can help you work effectively with systems integrators, consultants, and OSS vendors to identify the right system for your service needs.

Establish the operating procedures. Before you begin the vendor negotiation and contractual process, you will need to accomplish the following tasks in-house:

- *Develop business rules and processes.* They define how employees and information and support systems will work in your company. These business rules and processes are used to drive your OSS applications. Without first developing these rules and processes, any OSS system will find it impossible to support the organization because the rules and processes that define an organization are nonexistent. Without clearly defined business rules, a new company would be at the mercy of operating according to whatever applications it chose.

- *Create a paper model of the operating company.* Before embarking on defining requirements and selecting back-office systems, create a robust paper model of the operating company, detailing how the business will operate by functional area. This can be a long and exhaustive task, but the up-front effort pays off as you begin the next phase of vendor selection and implementation.

DEFINE A SET OF OSS REQUIREMENTS BASED ON THE OPERATING MODEL. This task can take several weeks; it can be done in house or outsourced. Ideally, defining functional and technical requirements for an OSS system comes from a detailed set of interviews involving the knowledge experts from every functional discipline in your company. You will need to interview customer support reps (CSRs) and service technicians, sales people, billing and collections, engineers and operations. This is a daunting task, particularly if you are not familiar with the process. If few formal processes exist in your company, and you are unsure of how to develop an OSS RFP, you may be better off hiring an OSS specialist who can help you do this. Most consulting agencies keep predeveloped RFPs that can be tailored to your operating model, thus eliminating the need to create one from scratch.

EVALUATE AND SELECT AN OSS VENDOR. If multiple vendors are going to be involved, you can consider involving a systems integrator to tie the different applications together. Alternatively, the prime vendor can undertake this responsibility. The vendor selection process itself involves a series of steps:

- *Develop a short list of OSS firms that have a reasonable chance of meeting the basic requirements within a specified time frame.* For example, the winning vendors should have one month to reach contract and four months from the signing of the contract to complete the implementation and turn the systems over to you.

- *Issue a detailed OSS RFP.* You may want to ask vendors to bid on an integrated solution and serve as the prime OSS vendor. In this case, the vendor will be responsible for developing consistent interfaces with any third-party applications needed to bid a fully integrated solution, and will also be responsible for integrating the different applications together.

- *Evaluate capabilities and develop a short list of two or three final candidates.* The evaluation process may include on-site demonstrations and detailed interviews of vendor references. Billing, provisioning, monitoring, and reporting capabilities are all key areas to examine when evaluating the capabilities of the OSS system. These are described in detail below.

  - *Billing.* The billing system generates invoices. The development of a superior billing system can be a substantial competitive advantage for you. Look for flexibility, so that you can accommodate the needs of different groups of customers. For example, small-business customers typically desire a bill that is simple to understand, whereas larger businesses (or specific groups of small businesses such as law firms) may desire a bill with more detailed usage and accounting information to track expenses or for cross-billing purposes.

  - *Pricing.* Pricing systems are used to set service pricing prior to service installation and are an input into the billing system that generates invoices. If your new service has a relatively simple pricing structure, you may not require a separate pricing system, but if your offering has multiple target markets, with different tiers of service and pricing structures for each, you may need to develop a more sophisticated pricing system. You should also make sure your system can accommodate unique events such as promotions, credits, and bundles of services.

– *Order entry.* Besides an initial sales call, order entry is often the first direct contact between a new customer and your company, so it is vital to make this experience as smooth and painless as possible. An order entry system that can accommodate all necessary customer information, for all components of the service, is a requirement. Experts recommend that BLECs, for example, maintain an easy-to-use Web portal for ordering services. Many data center providers have begun to migrate towards Web ordering interfaces, and this may be a possible interface (either as the primary ordering mechanism or as an adjunct to other methods like call centers) for service ordering. If a Web interface is used, it should be as simple and as user friendly as possible.

– *Provisioning.* Telecommunications services in general can be tricky to provision, because there are typically multiple providers involved in the process (e.g., ILEC, CLEC, ISP). Your service provisioning system should be well integrated with your order entry system, and it should also be well integrated with the provisioning systems of your partners (such as ILECs, IXCs, or network backbone providers). Some experts recommend that a true flow-through provisioning system should be a goal, because scaling a "swivel-chair" provisioning system, without true integration between the various elements, is difficult, if not impossible. Figure 6.5 shows the possible types of element interfaces that may be required for true flow-through provisioning.

– *Trouble ticketing.* Trouble ticketing can get complicated if you are offering services involving multiple networks and systems, such as integrated voice and data services. A problem with a "packet voice over broadband" service, for instance, could be on the data or DSL side of the network; it could be within the voice portion of the network; or it could, potentially, be in both segments of the network. It is important that your trouble ticketing systems be able to communicate with each other to facilitate rapid troubleshooting and to minimize duplicated efforts by your support personnel when tracking down a network problem.

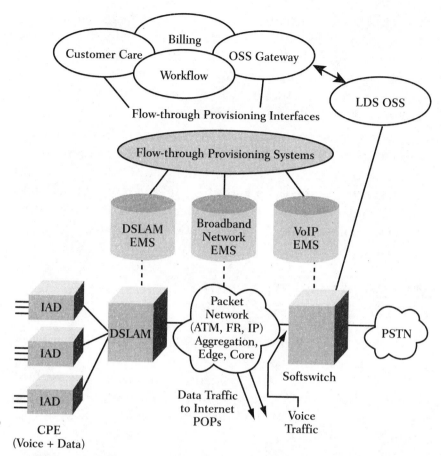

FIGURE 6.5   Flow-through provisioning interfaces.

— *Reporting capabilities.* Performance is one way to differentiate your services, and also forms the basis of your SLA. You will most likely be required to show network performance metrics as proof that you are meeting your performance goals. The best way of providing such data is to tie your network performance monitoring systems into Web-based reports that are visible to the customer. The reports delivered can be customized to meet the needs of customers (for example, simpler reports for small-business customers who don't need or want lots of technical data) and can be mapped to show performance metrics that correlate with your SLA.

- *Select the vendor who scores the highest and can reach a contractual agreement.* The final step involves selecting the vendor who scores the highest and can also work with you to meet implementation deadlines. You will need to negotiate pricing and contractual obligations; usually the systems integrator will facilitate these negotiations.

- *Test and pilot the system before deployment.* This is the final critical step in the process. You will need to identify test procedures and conduct a pilot test to ensure that your OSS is performing the tasks as specified. A pilot test is warranted to iron out kinks in the order entry, provisioning, and billing processes, and request improvements from the vendor if necessary.

METRICS FOR MEASURING PARTNERSHIP SUCCESS

If you can't measure it, you can't manage it. Having a tracking system in place helps you gauge your progress in implementing a new service with a partner, identify problems early on, and change direction swiftly before much damage is done. Some of the metrics that can help you measure the effectiveness of your partnerships include:

- *Churn rates.* How has the partner's service impacted your customer churn rate? Is it the same, lower, or higher? Churn rates indicate how successfully a company provides service. When you see a lot of churn, you see poor management, poor implementation, and poor systems. While it could be attributed to other factors, a higher churn rate since you started the partner's service may also signal potential problems with the service performance, support quality, and so on.

- *Value-added service revenues, as a percent of overall sales.* Has the partnership helped you achieve your revenue goals for the new service? If not, does the problem lie with the sales staff, the service features, or the quality of support provided by your partner? Could it perhaps be the perceived image of your partner?

- *Service outages and uptime.* Is your partner upholding its part of the SLA agreement?

- *Quality of technical support during network problems.* Is your partner addressing network issues in a timely and responsive manner?

- *Customer satisfaction with quality and reliability of service.* How are customers reacting to the new service from your partner? Has there been an escalation of customer complaints since the new service was deployed?

- *Depth and breadth of SLAs provided by the partner.* Is your partner providing the necessary assurances in the SLA to guarantee service reliability and quality? Are there gaps in the SLA that may signal lack of preparedness on the part of your partner?

PREPARING TO LAUNCH YOUR SERVICE

You must put together a well-defined launch plan before you begin to deploy a new service. This up-front planning will go a long way in providing insight into potential obstacles and challenges—whether financial, technical, or operational in nature—that could negatively impact your service rollout. You will want to know about these challenges and overcome them before you begin service deployment.

There are several steps to take to prepare for a full-blown deployment, some strategic and thought provoking, such as positioning, and some more tactical but equally important. These are discussed below, not necessarily in the order of priority, because most are done in parallel:

- Allocate roles and responsibilities for launch between corporate and organizational areas
- Flow out the timeline to deployment
- Start launch-related marketing and sales activities: service positioning and sales training followed by a marketing communications campaign
- Put in place necessary operations support and back-office processes

- Assess personnel needs and staff up as necessary in various organizational areas
- Conduct early customer trial

ALLOCATE ROLES AND RESPONSIBILITIES FOR LAUNCH BETWEEN CORPORATE AND ORGANIZATIONAL AREAS

THE FIRST RULE: INVOLVE YOUR ORGANIZATION

Involve the various functional areas of your organization in the details of the launch planning process. In particular, there are five key interdependent organizational areas that play a vital role in launching a new service: customer service and support, operations, planning, marketing and sales, and finance. Launching a new service, particularly a brand-new service, is a significant event for any service provider. Therefore, it is vital to gauge and prepare for the impact it will have on these different parts of your organization and on the positioning and marketing of existing services like data, traditional voice services, and Internet access. Assessing these impacts while you are planning your new service helps to minimize the effect on your organization and improve your odds of a successful product launch. Finally, the whole organization must mobilize to put the new service through its paces in the form of a market trial that could follow the launch or occur in parallel.

The key organizational areas that are involved in planning a service launch or rollout are shown in Table 7.1.

TIMELINE TO DEPLOYMENT

A careful examination of your current systems and capabilities (both financial and personnel) is required to evaluate how long it will actually take your company to deploy your new service. Table 7.2 shows a sample timeline for a BLEC to deploy a bun-

TABLE 7.1 Organizational Areas and Launch Planning Responsibilities in a Telecom Service Provider

Organizational Area	Activity
Planning	Putting together a requirements checklist for deployment
Marketing and Sales	Service positioning, sales training
Finance	ROI spreadsheet, personnel requirements
Early customer trial	Live network interoperability testing in lab followed by service trial with small customer base
Support	Back-office systems integration, provisioning, and OSS systems
Operations	Installation, on-site support, network readiness
Customer Service	Operator and help desk, training
Whole organization	Market trial

TABLE 7.2 The Total Elapsed Time for the Steps Below Can be as Little as Six Months if Done Concurrently

Step	Description	Time
1	Company decision to enter market	Zero days
2	Market research (e.g., understanding customer requirements and competition in your target market and evaluating competitive and substitute products)	Three to four weeks
3	Create market entry strategy	Six to eight weeks
4	Service plan definition	Four weeks
5	Billing structure and rate development	One week
6	Determine company and service positioning	One week
7	Installing network equipment and back-office systems	Four months to one year (you can shorten timeframe by doing workarounds before the system work is complete)
8	Create launch plan	Four weeks
9	Create market messages and Marcom Plan	Four weeks
10	Test marketing	Four weeks
11	Launch service	As soon as all critical steps above have been accomplished

dled voice and data service, based on the assumption that the service offering includes high-speed Internet access and local and long distance analog voice services, and that the resources and financing are in place for the initial deployment. Note that some of these steps can be conducted in parallel, reducing the overall time to market, if sufficient resources exist to do so.

LAUNCH-RELATED MARKETING AND SALES ACTIVITIES

Positioning activity is an important part of the launch process and forms the basis for your marketing communications campaign (the latter is the subject of Chapter 8). Although this section deals primarily with service positioning, and only touches briefly on company positioning, it is important to realize the difference between service and corporate positioning. Too many marketing teams rush to create positioning documents during launch time without understanding what it is they are really setting out to do, and producing something that addresses neither service attributes nor company attributes.

A good rule to remember is that company positioning generally leads in establishing market presence or reach, and service positioning generally leads in establishing product differentiation. Different ways to establish effective market presence were discussed in Chapter 3. Company positioning is thus part of the process of developing a company image and articulating what it is you are best at to your customers. This positioning is an exercise that looks inside your company's deepest competencies and explores what makes your company and services shine or stand out to your customers—your compelling value proposition. These attributes are then captured to create a set of unique positioning statements that become the basis for all corporate and customer communications until they are firmly entrenched in the market's mind over time.

Beyond the compelling value proposition, company positioning can be supported by attributes such as your business goals, revenue numbers, comprehensive service offerings, market share or leadership position, network completion, marketplace endorsements, top-tier customers, and so on. The three key influencers of your company positioning exercise are your value proposition, your target market needs, and the resultant opportunity, and your business goals.

While retaining the key elements of company positioning through subtle or explicit messaging, service positioning is supported by service-centric attributes such as cost savings, network presence, quality of service, service level agreements, service features, competitive advantages and so on. The most ambitious goal of service positioning is to portray you in a light that uniquely differentiates you from the competitive herd, encourages industry experts to endorse you, partners to want to partner with you, and customers to buy from you.

SERVICE POSITIONING GUIDELINES

When positioning your service, remember that you are also reinforcing company positioning. Positioning is more than the sales document that you put together to train your sales force and to make customers aware of your service offerings. Positioning is also more than a feature–benefits statement, which is really a fact sheet. Positioning is, instead, a powerful vehicle for communicating your corporate and service value propositions to the marketplace. The next sections explore some key things to keep in mind when positioning your service, whether to a new market or to an existing customer base.

SERVICE POSITIONING INFLUENCERS

PRICING AND POSITIONING: TWO SIDES OF THE SAME COIN. When positioning your new service, refer to your pricing objectives, because they will have a significant impact on how you position your service, and vice versa. For example, is your business goal to achieve rapid, widespread adoption of the new service? The positioning you put in place, accompanied by the appropriate pricing strategy, will be the key components of implementing this strategy. Pricing and positioning are closely linked. In fact, price can be a vehicle for positioning your service and is frequently viewed by customers as a positioning statement about the type of company you are and the level of services you provide.

OTHER IMPORTANT INFLUENCERS.   The primary influencer of your positioning is your target market needs. What do prospective customers perceive their needs to be? What matters most to them—quality, price, support, or product? What services are they currently buying, and from whom? It also helps to understand your customers' profile: How technically savvy are they? For example, small businesses are less likely to be swayed by the technical merits of a VPN service and focus more on the price savings they get by eliminating long distance calls. By contrast, medium-sized businesses have some measure of IT expertise and are more likely to focus on the technology and integration issues before they consider the price savings. Your message to a small business would therefore focus more on price savings and ease of use, and on ease of integration and leading edge technology when targeting medium businesses.

Other important influencing factors are listed here:

- *Service positioning must align with your corporate goals.* For example, if growing market share quickly is one of your corporate objectives, then you want to make sure you are positioning the service to achieve this objective. If your company goal is X, both your pricing and positioning must support this goal.
- *Your service positioning must also align with your core value proposition.* What is your company known for being the best at? What is the compelling reason that customers buy from you? How does your new service fit into your value proposition; through features, quality of service, network presence, or some other form? Emphasize this so that customers get the message. Your new service must reflect your value proposition or you stand the risk of either confusing customers about what your value proposition is, or creating the impression that the new service is not a core offering built around what

your company excels at. One way to align the service positioning more closely is to update the key messages and statements surrounding your value proposition to embrace the key attributes of your new service, thus bringing it under the umbrella of your value proposition.

· *Service positioning is also influenced by industry trends, whether of a regulatory, technological, or competitive nature.* Industry trends can have an effect on customers' buying habits and attitudes and impact how customers view new telecommunications services. Industry trends can also impact the competitive landscape and pricing pressures, thus affecting how you position your service. The most pervasive trends to watch for in the telecom industry are of a regulatory or technological nature. Some of the things you may want to watch for include government intervention in the local phone markets, the introduction of new technologies or regulations that make fiber deployment a near-term reality, and the impact this is likely to have on customers in your target segment.

· *The competition in your marketspace will also influence your service positioning.* How is your competition positioning itself? Are there potential weaknesses or gaps that you can exploit? Positioning charts and graphs often help. Plot the key attributes, such as value, brand recognition, or target customers, for example, and place your company and your key competitors on it. Are your competitors positioning the same way as you? This exercise can help you focus on positioning your service in a way that exposes weaknesses in your competitors' offerings.

At any rate, don't underestimate the competition. Be on the lookout for other providers that are attempting to address the same space with a competing technology or by dramatically lowering prices. To help you internally differentiate and position versus the competition, the template in Table 7.3 can be used to identify some of the key characteristics of your service versus the competition. You can add to this other specific features that the market considers valuable and may further differentiate you from your competitors.

TABLE 7.3 A Template for Positioning Your Service versus Your
Competitor's

POSITIONING CRITERIA	YOUR SERVICE	THE COMPETITOR'S SERVICE
Market Positioning		
Pricing		
Reliability		
Service Assurance (SLAs)		
Flexibility (truck rolls, configuration)		
Scalability		
Customer Acceptance		

- *The unique features of your service deserve their place in service positioning.* You will want to highlight these features from the standpoint of how they benefit customers, build on your company's core strengths, expose weaknesses or gaps in competitive offerings, highlight a service innovation or "first," or demonstrate excellence in some other way that is beneficial to customers.

- *Market dynamics will shape your positioning as well.* Are you in the midst of an economic slowdown? How can you tailor your service positioning to reflect customers' changed attitudes about buying?

CROSS-POSITIONING WITH EXISTING SERVICES

If you are already in the market, service positioning becomes an essential activity that helps your organization and the marketplace understand where the new service fits into your existing portfolio, how it meets customer needs, and when and why customers should buy this service. If you are already in the telecom service business, then the new service will need to be "cross-positioned" with respect to the other services you offer. For example, if you already offer Internet access and are planning to offer a full-fledged managed service such as dedicated Internet access with VPN and Web hosting, key questions your sales force will want answered are: will the new service be positioned as an upgrade (e.g., secure "always on" Internet access) or future value-added service for basic Internet access customers, and if so, will this affect which service level they should choose? Customers will need to be educated on the benefits of the new service and why they should upgrade to the new service. This is done via cross-positioning statements created by the marketing team.

The cross-positioning exercise should output a somewhat formal stance in the form of a sales document that can help your sales team determine which customer segments should be guided towards the new service, and which should be guided towards those other of your existing services that make more sense for them. In parallel, the marketing team will also need to lay out a set of customer criteria that can help your sales force identify prime candidates for (upgrading to) the new service. Examples of what these criteria could be are size of the business, type of business, branch office locations, annual revenues, size of in-house IT staff, and so on. This group can then be targeted with an upgrade incentive program if necessary.

If you are undertaking a cross-positioning exercise, then here are some guidelines for what you will need to accomplish:

- A matrix of which services to actively sell or emphasize (e.g., Internet access, managed VPN, etc.) in specific customer situations, along with the profile of the target customer to whom you sell each service.

- A migration strategy, if applicable, for moving customers to the new service, and the financial implications (via business case modeling) of this migration. The financial impact must be considered, because it is quite possible that the new service will have lower rates and may bring in lower revenues than existing services.

- Promotional incentives (discounted bundles, rebates, other types of promotional pricing, etc.) to accelerate take rates of the new service.

- Determine the kinds of customer behavior (e.g., usage patterns, contract length, value-added service purchasing) you wish to encourage for each service. Mapping customer characteristics to particular services, based on each service's unique characteristics, can help clarify how each is positioned vis-à-vis your customer base.

A side benefit to the above cross-positioning exercise is that it helps you think strategically about how the different services in your overall product portfolio interrelate and impact each other. For example, it allows you to:

· Gain a clearer understanding throughout the organization of the benefits and weaknesses of a new solution compared to existing services

· Leverage your efforts by creating customer checklists that can be used as sales tools to help your sales team pre-qualify specific groups of customers for the new service

· Evaluate revenue, profitability, and market metrics for all of your services, which will help you to determine successful strategies for all or affirm existing strategies

· Coordinate strategic business directions, messaging, and other corporate intangibles across your product portfolio

· Provide feedback to your pricing strategy, in terms of both pricing structure and actual rates, so that you can target those customers that best fit your service offering

Conducting cross positioning is not always an easy task. You may find, for example, that "competing" product management teams within your company will have conflicting positioning desires for their respective products, especially if an existing service stands to lose revenue to the new service. Involvement of upper-management decision makers, who have the strategic vision and authority to mediate in such cases, is essential to a successful process.

OPERATIONS SUPPORT AND BACK-OFFICE PROCESSES

Launch planning includes implementing the necessary internal operational support structure and setting up your internal organization to manage the customer-support process, from taking the order and requesting the service, to servicing the customer on an ongoing basis. Although a vital component of service rollout, this is easier said than done by most start-up carriers. Although every telco aims to provide great customer support, the diligence required to put in place the necessary procedures and support staff is usually lacking, especially if management gets so involved in the technology and marketing aspect of the launch that very little is done to shore up customer support.

Experts agree that most CLECs completely underestimate the important role quality and diligence play in servicing the customer. More specifically, failure to map processes properly and assign responsibilities down the chain can lead to different teams focusing on their own areas and not understanding where and how they fit into the big picture. This causes a communication void that results in chaos. The business plan is seldom shared with the support organization, and the focus on building support procedures is minimal because the customer base is young, orders have not started coming in yet, and it is relatively easy to service a small group of customers with entry-level processes. The theory is that the CLEC will easily be able to expand as the volume grows. The Management Network Group's Marsh notes that whether you initially manage customer support using spreadsheets or a series of entry-level applications, you have two things to think about. What if you grow beyond the entry-level application? How will you manage the support process today and in the future?

Of the two, managing the customer support process is more important. It is not just taking the order and requesting a service, it is the entire process that performs the following functions:

· Accepting the order
· Validating the customer
· Validating the service requested
· Requesting the services internally

- Requesting the supporting services externally
- Assuring that the service is set up properly and in a timely manner
- Maintaining contact with the customer
- Validating that the service works
- Verifying that the billing components are set up and rates are correct
- Verifying and processing daily usage
- Verifying the billing
- Processing payments accurately and in a timely manner
- Managing vendor costs
- Processing and auditing commissions
- Ongoing servicing of the customer
- Handling the exceptions to all of the above

Defining a well-organized support chain in the call center or support help desk can take time. Even before the first customer signs up, you will need to flow the processes to a detail level. Start at a general level and lay out customer entry points, basic qualifications, how orders should be written, the flow of an order through the provisioning process, and the handling and installation of the service in either a resale or facility mode. Other factors to consider include account setup and billing management, and methods of servicing customers. How do you expect to manage vendor costs? And, most important, how will you deal with exceptions along the way?

The other key component in this area is ensuring your back-office and OSS systems are functioning and have successfully gone through the pilot testing process, as discussed in detail in Chapter 6. In general, deploying your operational support systems, integrating network components, and getting ready for the operational side of the launch requires a dedicated team leader who can both define and oversee the deployment process from network design to launch (Figure 7.1).

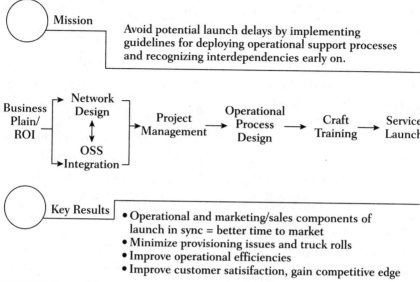

FIGURE 7.1   Components of a well-planned launch.

PERSONNEL REQUIREMENTS

Launching your new service may well be your first entry into the world of telecom services. If so, then this is a move into unfamiliar territory, and it is a good idea to evaluate carefully your in-house expertise and the skills of your personnel and review the capabilities of your back-office systems to ensure a smooth service launch. Don't underestimate the importance of this task—many promising and technically feasible new services have been delayed or launched unsuccessfully because the customer care elements, like provisioning and billing, were not properly prepared to support the new service. We discussed the process for identifying your back-office requirements and selecting the appropriate OSS systems based on these requirements in the last chapter. Thus this chapter does not deal with the back-office equipment selection and setup, but instead provides insight into the process for assessing your staffing needs in the various organizational areas—another critical component of a successful launch. You may want to use staffing metrics that help you identify gaps, close these gaps, and outperform the competition. The very first thing, however, is to understand the expertise requirements for the various departments. Are you going to be staffing from the ground up or incrementally?

Training is another area that can exhaust valuable resources during launch time. You will want to consider the training demands of the new service launch on your organization. How many new personnel will need to be hired to accomplish this? What kinds of new skills, knowledge, or expertise will your sales engineers and service teams require, and what will be the impact on their current assignments to provide this training?

A good place to start is to answer the following question: are you going to be staffing from the ground up or incrementally? For example, if you are launching an integrated voice and data service, and you already offer analog (POTS) voice, then you may well find that your personnel already have most of the skills necessary to support the launch of broadband voice, thus making your entry into this new market easier and faster. If you are a brand-new entity launching integrated services, however, then you will need to staff appropriate areas from the ground up versus incrementally to support the launch of your new service. Recognize that in today's financing environment, the cash burn necessary to support these areas, particularly support, billing, and provisioning, can be intensive. Key to addressing this issue is the procurement of qualified talent and the management of that talent.

For example, some of the critical skills necessary in your organization to support an integrated voice and data service offering may include knowledge or expertise in:

- Traditional voice services such as analog voice (PSTN, POTS)
- Data networking services in general and broadband services in particular
- CPE, such as integrated access device (IAD) or firewall/VPN appliance
- Target market (typically small to medium business) voice systems such as PBX, key phone systems, fax, and analog handsets

- Customer support and provisioning systems
- Selling voice and/or data services

These skills will likely be divided among several different teams within your organization, such as:

- Executive management
- Sales organization
- Back office and support
- Operations organization
- Customer service
- Planning organization

EXECUTIVE MANAGEMENT

First and foremost, you must assess the completeness of, and if necessary recruit into, your core management team. To do this, you must know what management talent is available in house and what you may need to bring in from the outside to augment your existing talent pool. You must analyze both the available expertise and the required expertise. This includes knowing:

- The skill set for each executive position available or required
- The proposed structure for managing the company
- The timeline for building the infrastructure to support the company

SALES ORGANIZATION

Depending on the type of service you are launching, you may choose to sell your service offering through one or more of three sales channels: direct, agent, or reseller. Regardless of which channel you use, your sales team should have the ability to convey the business advantages of your service, not just the technical ones. Remember that you are selling a service, not a technology!

For instance, to be able to communicate the benefits of a somewhat advanced telecommunications service that includes broadband voice and managed security services, your sales team must be knowledgeable in the following areas:

· Telecom needs of your target customers
· Traditional business voice services (POTS, Centrex, fax)
· Business telephone systems (key systems, PBX, analog phone systems)
· Data services (Frame Relay, ATM, IP/VPN, Ethernet WAN)
· Access services (DSL, T-1/E-1, fiber, Ethernet)
· Security services (firewall, VPN, anti-virus, intrusion detection, authentication, encryption)
· Telecommunications market trends and emerging services

Your new service will most likely target both small- and medium-sized businesses and may possibly target larger businesses for certain applications, such as branch office connectivity and secure telecommuting. For the most part, your sales team will require similar skill sets for all of these potential customers. However, it is important to realize that when selling to larger customers, it is likely that the decision makers (generally IT staff) will be more technical, so your sales team may be required to have a higher degree of technical knowledge about the system.

BACK-OFFICE AND SUPPORT STAFF

Back-office and customer support personnel are not typically in the spotlight on product rollouts, but they play a vital customer-facing role in your service provisioning process. Indeed, in many cases—particularly for small-business customers—these employees may be the first face-to-face interface between your company and a customer. If you are offering integrated broadband services, then these personnel must be trained in a wider array of areas than those who are selling only voice or only data services, because integrated services involve multiple services (e.g., voice, data, and Internet access) over a single network, therefore creating a more complicated network.

This team should have skills in the following areas:

- Basic knowledge of both voice and broadband services
- Ability to enter customer orders, track problems, and suggest solutions
- Customer service skills
- Ability to verify receipt of all necessary customer information
- Knowledge of value-added services (such as security services or SLAs) in order to upsell services
- Ability to forward information to provisioning personnel to begin the installation process

OPERATIONS ORGANIZATION

Your operations organization is critical to your new service. Provisioning and other operational activities have proved to be one of the most difficult areas for current broadband service providers. The addition of voice services to data makes the operations organization even more important. The operations organization should be able to:

- Install and test network and customer premises equipment
- Understand common business phone equipment
- Monitor and track network performance to ensure service levels meet agreed-upon metrics
- Manage and monitor network connections with partners (such as ISPs, IXCs, or ILECs)
- Troubleshoot service problems
- Perform fraud detection and prevention

CUSTOMER SERVICE AND OPERATOR ASSISTANCE STAFF

Once customers have been acquired and have had service provisioned, much of their ongoing interface with your company will be with your customer service and operator assistance groups. If you now offer a traditional voice service, these groups should already be well equipped to perform their required tasks. Many service providers, particularly those without existing voice service portfolios, choose to outsource this particular function. Your customer service and operator assistance groups should have the ability to:

· Address customer billing inquiries
· Initiate trouble tickets and provide initial troubleshooting
· Handle operator assistance and information (411) calls
· Initiate basic testing of customer voice lines
· Answer basic customer questions about services

PLANNING ORGANIZATION

Your planning group is responsible for engineering and designing your network and also for planning network expansion as your customer base or service area expands. The planning group should consist of both engineering personnel and product management personnel to ensure that both engineering and business considerations are properly addressed. Among the knowledge areas required by a planning group are:

· PSTN (network infrastructure, SS7, etc.)
· Backbone transport technologies
· Broadband technologies (DSL, T1, wireless, fiber)
· CPE technologies (IADs, appliances)
· Business model understanding (ROI, margin, depreciation, etc.)
· Telecommunications market trends
· Telecommunications protocols
· Emerging alternative or complementary services (IP telephony, etc.)

EARLY CUSTOMER TRIALS

When a trial begins, 90 percent of the service planning should be in place. The purpose of a trial is to validate service acceptance and iron out kinks in the provisioning and rollout process, not to help you figure out if you chose the right technology or access solution for your service rollout. A trial's resources should not be wasted on obtaining technology validation from customers; instead technology validation should be taken care of beforehand through a technology trial conducted by the equipment vendor in the service provider's lab or in the vendor's own interoperability lab. As we have reiterated in the chapters preceding this one, customers are less focused on the technology per se, and more concerned only that it works. Service provider trials should essentially focus on the marketing, sales, operational, and support aspects of the service, and minimally on the technology of the service.

MARKETING COMMUNICATIONS— MAXIMIZING YOUR MARKETING DOLLARS

The market for small and medium business telecommunications services is crowded with "me-too" products of all sorts, so having a clearly defined and effective marketing communications strategy is a key means of differentiating your telecommunications service. Your marketing communications program is, more than anything else, a way of defining yourself in the marketplace, and it has a direct influence on how potential customers and others (such as market influencers like analysts and press) perceive your company and your services. While simply being first to market in your target segment with an exciting new service may get you some mind share to begin with, it is no substitute for an ongoing, targeted messaging program. There are a few simple steps that will get you started on building a set of messages that will form the basis for your marketing campaign.

STEP 1: RECAP YOUR SERVICE OBJECTIVES

The first step in developing your marketing communications (marcom) strategy is to understand clearly your corporate and service strategy. Your corporate strategy drives your service strategy, and your service strategy drives your marcom strategy. Therefore, it is imperative that you understand exactly what it is you hope to accomplish from launching the new service. For example, your service strategy may call for achieving the following objectives:

- Increasing your overall market share among small and medium businesses
- Attracting new customers within your service area
- Retaining existing customers
- Providing new value-added services to upsell existing customers
- Differentiating your portfolio of services from those of competitors

Identifying these objectives helps you conduct a proper evaluation of the strengths and weaknesses of both your company and your new service offer. It also facilitates your evaluation of these characteristics in comparison with both market needs and competitive offers.

STEP 2: RECAP MESSAGING DRIVERS

Preparing your key messages and crafting your marketing campaign is simple if you have already completed your internal and market assessment, your company and service positioning is in place, and you understand your relative strengths and weaknesses versus those of your key competitors in your chosen space. If these steps have already been completed, you can now leverage the results of this previous work.

Performing these evaluations enables you to create a set of compelling high-level positioning statements and messages about your service, which you can then prioritize and use to create more specific marcom messages. These high-level positioning statements become the basis of your ongoing marketing messages, and they enable you to focus all of your marketing efforts in support of your overall strategy.

The following are a few things to consider as you proceed through the marketing communications development process:

- Consider your overall corporate strategy and objectives.
- Conduct a thorough market research assessment of your competitors and their messages so that you can avoid giving the impression of a "me-too" player in your target market.
- Leverage the expertise of marketing and PR resources, both internal and external. For example, use your ad or PR agency, and internal PR staff. They probably have already gone through exercises like this for other products.
- Be farsighted. Focus not only on messages that are compelling today but also on messages that will be effective six to eighteen months down the road.

STEP 3: CREATE POSITIONING STATEMENTS

After analyzing the various drivers behind your messaging process, it is time to create your high-level image or positioning statements. These statements are designed to introduce your company and your new service to potential customers.

Positioning statements provide the basis for a customer's first impression of you and your service and should be specifically targeted to their audience. Therefore, you may have more than one set of initial positioning messages. For example, small business may be focusing primarily on the potential cost savings of DSL, while medium to large businesses may react better to a message that emphasizes the service's ability economically to provide the same kind of service to branch offices that large offices currently enjoy. Table 8.1 presents some sample positioning statements for a converged voice and data service to two different customer groups.

TABLE 8.1   Positioning to Different Customer Groups

Target: Small Business	Target: Medium to Large Business
One-stop shopping for all your communications needs from a new type of provider	Leading-edge integrated communications services for all of your offices and telecommuters
Full voice services AND high-speed Internet for less money	Easy to integrate with existing phone and communications systems
A single provider who can meet all of your communications needs	Flexible, upgradeable service over a single access line

Here are a few final points to consider when crafting your initial positioning statements:

· How do your proposed service positioning statements compare to your company's existing image and positioning? Do they enhance it?

· Is your existing image weak, and if so, should your new messages ignore it to improve your overall positioning?

STEP 4: EMBED YOUR VALUE PROPOSITION IN YOUR MESSAGE

The next step in the process is to create a series of statements around your value proposition—statements that clearly demonstrate to customers what value your service offers to their business. Again, these messages must be targeted to the individual needs of each of your target markets to have the greatest impact on each group.

It is important to note that the value messages you create here really define who you are as a service provider and should be used both internally and externally. Getting all of the internal stakeholders (like sales teams, customer service, and marketing) on the same page is key, so use your value propositions for internal training and documentation as well as for outward-facing marketing communications.

Your value proposition messages may be the first time that customers have heard about your new service, especially if it is a brand-new service not yet being offered by the competition, for example unified messaging or packet voice. It then becomes important to elicit some response from potential customers. To do this, you must focus on exactly what value the new service offers to a customer's specific needs. For example, for small-business customers, you may wish to focus on the cost savings that a packet voice solution provides compared to traditional multiline POTS or Centrex service. You may also want to emphasize the value of utilizing a single service provider for both voice and data services using related administrative overhead reductions. In the case of converged services, another value proposition for a small business may be to call out the flexibility of your service plus the ability to quickly provision additional lines when a customer's needs change.

You may wish to create your value propositions for a market (like small businesses) by addressing three specific areas: hot topics for that segment, current business problems, and customer education to overcome concerns about a new technology. Using broadband voice and data services as an example, Table 8.2 shows sample value propositions for the small business market.

These examples provide a good starting point for creating top-level value messages that will interest and excite your target market customers. These general messages should be combined into concise marketing statements. For example, one such message may be "Broadband voice leverages your high-speed DSL connection to provide multiple lines of voice." Another example might be "Broadband voice maintains the quality of traditional voice services, while lowering costs by using your network connection more efficiently." Or, "Broadband voice provides you with both voice service and high-speed Internet without your having to deal with multiple service providers." The key is to create statements that are broad enough to allow you to create supporting value statements for each but specific enough to grab the attention of a potential customer.

TABLE 8.2   Sample Value Propositions around Converged Services

SMALL BUSINESS HOT TOPICS	SMALL BUSINESS CHALLENGES	TECHNOLOGY EDUCATION
Convergence of voice and data onto a single network is the future of business access.	Telecommunications costs, especially voice, are high and affect profitability.	Does broadband voice service offer the same quality as traditional voice services?
Converged broadband services give small businesses the same level of service enjoyed by large businesses.	Internal IT and administrative resources are scarce and internal expertise is limited.	How does a converged service save money by more efficiently utilizing broadband connections?
Broadband service leverages the high-speed broadband connections being adopted by most small businesses.	What kind of productivity increases do new services and technologies like VoIP enable?	Does broadband voice service have the same level of reliability and lifeline support as traditional services?

CREATING A MESSAGE HIERARCHY

The most effective messaging is that which starts out with a set of top-level messages, and then drills down with a set of supporting messages behind each top-level message. Having a series of progressively "deeper" supporting messages applies focus to your messaging and ensures that all of your supporting messages clearly support your main messages. Too often, new services are launched with an "all-things-to-all-people" message, creating confusion in target customers. A hierarchy allows you to drill down to the key value messages for your customers and to maintain a consistent message in all of your marketing communications.

TeleChoice recommends creating a pyramid that begins with an overarching high-level statement that introduces your company, your new service, and its value to the customer. Like all of the other components of your messaging strategy, this high-level statement is customized for each target market, so that your small-business pyramid may differ slightly from your medium- to large-business pyramid. TeleChoice recommends creating different pyramids for other groups, like the press or the analyst community, to tailor your message for each group.

In the TeleChoice pyramid shown in Figure 8.1, only three second-level statements are used to support your top-level statement and answer the question "What's new or different about this service?" Each of these supporting statements is in turn supported by three more supporting statements. The third-level statements are more specific and detailed than the level above and provide direct support to the second-level statement that precedes it. The third-level statements form the support of the messages you send out to customers in the form of press releases, brochures, Web sites, advertising, and the like.

Using the messaging pyramid scheme, Table 8.3 provides a set of sample messages for a converged voice and data service targeted to small-business customers.

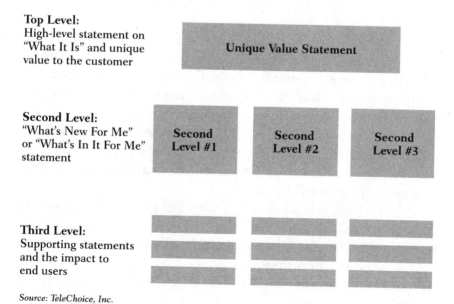

Top Level:
High-level statement on "What It Is" and unique value to the customer

Unique Value Statement

Second Level:
"What's New For Me" or "What's In It For Me" statement

Second Level #1   **Second Level #2**   **Second Level #3**

Third Level:
Supporting statements and the impact to end users

Source: TeleChoice, Inc.

FIGURE 8.1   Messaging pyramid.

TABLE 8.3   Sample Messages Based on the Messaging Pyramid

TOP LEVEL	THE SIMPLE VOICE AND DATA CHOICE FOR SMALL BUSINESSES		
Second-Level Statements	Converged services reduce voice costs	Same quality as traditional voice services	Simpler for SMBs to implement
Third-Level Statements	Utilize a single access line to get up to eight voice lines	Technology utilizes bandwidth prioritization to ensure toll quality	All data and voice services provided over a single line and through a single piece of equipment
	Uses broadband network connection most efficiently	Broadband voice equipment interfaces with the PSTN to provide the same connectivity as traditional services	Business customers can interface with a single point of contact for all their telecommunications needs
	Converged services lower equipment and network management costs—only one network to maintain	[Service Provider] offers service level agreements to guarantee service quality	Upgrade service quickly without new installations or equipment

If at all possible, you should try to test your messages with five to ten potential customers, current customers, industry influencers, or some combination thereof. Message testing is relatively simple and can provide extremely valuable insights into how the market will receive your message. It is much better to test it before you launch and make necessary adjustments, if needed, rather than find out something is wrong after you have printed your marketing materials and spent time and money on an advertising campaign.

Ask the test respondents if the message and supporting statements are believable (do they think you can deliver?). Ask them if the message is compelling (will they consider your company now, when they would not have in the past?). Give yourself enough time when conducting these interviews to be able to tweak the message if necessary.

STEP 5: GET READY FOR THE WHOLE CAMPAIGN

You have prepared your service definition; planned and organized all the marketing, operational, and support aspects of launching your service; and you are now ready to launch your service. How do you go about getting mind share from your target customer? In other words, what is your whole marketing campaign going to look like? Have you created your marketing communications strategy or mission? What are the different media you are going to use to reach customers with your marketing message? Depending on your budget and market size, you may want to use the Web, advertising, white papers, direct sales calls, and any other number of ways to market to your customers. Is your marketing campaign going to consist of highly focused one-on-one marketing with your customers, or are you planning to institute a media blitz with guerrilla tactics? If you are a BLEC serving a well-defined set of buildings in a metro area, a highly targeted marketing campaign may be your best bet. On the other hand, if you are serving a broader market, you may want to invest in some variation of a media blitz for maximum market exposure. You should already know the answer to these questions, because you gave them careful attention during your planning sessions. Regardless of the media you use to communicate your messages, a successful marketing campaign must revolve around the following key points:

· Capture your customer's attention
· Communicate clear benefits of the service
· Follow with incentives to buy

One of the key things your campaign should focus on is capturing your customers' attention. You will need to know what makes the customer "tick," and what are their hot buttons. Profiling the customer in the early stages of planning is an excellent way to determine not only what your target customer's needs are, but also to understand hot buttons. This helps you translate your service features into benefits that address those hot buttons, so that customers pay attention. Of course, you can use gimmicks to capture your customers' attention initially, but these are useless unless you also address their needs.

Translating the features of the service into benefits gained by the customer is the first step to laying out a marketing communications campaign. Keep in mind that not all service features need to be translated into benefits. You will not need a feature–benefit translation for basic features, such as caller ID, which are well understood. But if part of the value you are offering is different combinations of features or particular uses of your service, then make sure that the customer benefits are well understood. The feature–benefit process can also include an education step, particularly if you are launching a service that is not well understood or one that requires the customer to participate on an ongoing basis beyond the purchase. An example of such a service is a managed service.

To expand on these key points, let's walk through an example. Let us assume you are a managed services provider, and you are going to develop an ongoing revenue stream based on providing and managing, along with voice, a suite of Internet data and security services needed by small- and medium-sized businesses. You have determined your target customer need, and mapped your service offering to meet those needs. Accordingly, your bundled service offering includes, but is not necessarily limited to, Internet access with e-mail, Web hosting, a managed firewall, and VPN. You may decide to include POTS (local and long distance) voice, thus offering a "one-stop-shop" solution to your customers.

In this case, your marketing mission will be to persuade the customer to give you the outsourcing business, and you can accomplish this by educating the customer on the benefits of your service approach, specifically the implementation issues and the benefits of outsourcing. This mission—customer education and the resulting awareness in a preset sales target—will be the basis of your marketing campaign. Depending on how differentiated your service is, and what the competitive landscape looks like, your mission may also include an aggressive approach to gain market share (by the use of discounts and other sales promotions).

In evaluating your new managed service offering, your customer (a small- to mid-sized business) is going to have two basic choices:

· Do it all in house, assembling the necessary know how, purchasing the equipment, and hiring the staff needed to set things up and keep them running, or

· Outsource part or all of it to a Managed Service Provider such as yourself.

Your marketing campaign, in this instance, will build from a close examination of the implementation issues encountered by customers in undertaking the task themselves, followed by the benefits of outsourcing to you. An example of how you go about doing this is described in the next section.

THE EDUCATION OF COMMUNICATING BENEFITS

EDUCATE THE CUSTOMER ON IMPLEMENTATION ISSUES. Some of the implementation issues to educate customers on may include highlighting particular factors. Setting up and maintaining the various services (Web hosting, firewall, VPN, e-mail) in a business involves much more than simply buying and installing the equipment. A customer wants these services to be reliable and dependable without incurring huge up-front costs. Further, point out that a typical Internet service would require the customer to undertake a myriad of planning and administration tasks and require substantial in-house resources, such as:

· Planning
· Purchasing, installing, and configuring equipment
· Monitoring

- Troubleshooting
- Reporting
- Analyzing
- Upgrading

Along the same lines, a security service would require the customer to undertake:

- Generating a security policy as part of the planning
- Selecting a firewall product and implementing firewall rules as part of configuration
- Verifying continued security using active intrusion detection tools and verifying performance as part of monitoring
- Detecting intrusion attempts
- Generating summary reports of access patterns
- Analyzing logs

And a Web hosting service would require the customer to undertake the following:

- Initial capacity planning
- Site design
- Performance monitoring
- Analysis of traffic patterns
- Access logging and reporting

Similar considerations apply to other services, such as VPN or e-mail.

DESCRIBE THE RISKS OF IN-HOUSE IMPLEMENTATION.   Such management tasks may be more than the customer is willing to take on, for the following reasons:

- A staff of information technology (IT) experts is needed; this is a resource sink for a small company that would rather put its resources to work in non-IT areas.

- Risk exists in doing this for the first time, in the sense that all contingencies cannot be anticipated and schedules are not predictable. An analogy might be attempting to be one's own general contractor to build a house.
- Costs prove difficult to predict and control.
- The customer takes on the full responsibility for success or failure.

ARTICULATE THE BENEFITS OF OUTSOURCING.   By contrast, you should stress that significant benefits accrue to the customers' outsourcing Internet and data service to you:

- This permits the customer to concentrate on its core business, not on IT.
- Costs are lower, especially for complex functions that require 24/7 coverage.
- The outcome is more predictable.
- You, the service provider, are accountable, and you have a vested interest in maintaining your reputation.
- Outsourcing makes more sense for services that require more expertise or up-front investment. Firewall, VPN, Web hosting, and e-mail are all good candidates for outsourcing.

PRESENT THE BENEFITS OF YOUR MANAGED SERVICE.   Your feature–benefits statements for the example managed service would look something like Table 8.4. (Such statements can be as elaborate or as simple as your company preferences dictate; the size and format are irrelevant as long as the point gets across.)

USING INCENTIVE PROGRAMS

Discounts and promotions are both incentive programs that serve key purposes in the introduction and life cycle of a service. Discounts are often used to reward or encourage desired customer behaviors. Promotions are typically used to generate customer leads and accelerate service adoption.

TABLE 8.4   Sample Feature-Benefits Statement for a Managed Service

SERVICE FEATURE	CUSTOMER BENEFIT
Local and long distance calling plan (includes x minutes of long distance calling at flat rate)	A single bill from a single provider is convenient and makes financial planning easier.
Ten email accounts with Internet access for a flat monthly fee	Key service options are bundled to provide customers with a cost-effective package for their standard needs.
Web site hosting and domain name service	Business is not interrupted by an outage; customer does not have to maintain expensive on-site Web servers or staff a Web administrator/designer. This results in lower total cost of ownership.
Complete firewall security for all business transactions, secure business-to-business connectivity, and secure remote access for telecommuters and remote employees	All business transactions and communications are handled with complete security, so that customer's data is kept safe. The MSP handles maintenance, upgrades, etc. Customers do not have to worry about technology obsolescence.
Add users quickly	Customers can add a new e-mail account or user within 24 hours. This provides flexibility to add new accounts for subcontractors, visitors, or new employees on a quick, as-needed basis.
SLAs	Customer holds you accountable for acceptable performance.
Complete outsourced IT solution for a monthly fee with an annual contract	The customer gets a single point of contact for all design, piloting, deployment, help desk, and technology management services. This eliminates the need to coordinate different vendors and monitor the system in house. A managed environment provides redundancy, security, cost savings, and convenient billing, without any up-front investment.

Some discounting options include:

- *Volume discounts.* These are usually based on dollars per month; for example, a $10K level of use receives a 5 percent discount.
- *Term discounts.* These reward customers who sign long-term contracts. Higher discounts are offered for longer contracts (such as three-, four-, and five-year terms).
- *Affinity plans.* These plans tie users to particular calling groups or offer additional incentives for using the service. Friends & Family® is a famous voice plan example. Other affinity programs include earning airline miles for every *x* number of minutes.
- *Service bundle discounts.* This approach applies a discount when customers purchase additional services, such as ordering voice or other data services. This can encourage customers to adopt new services that you offer.

Many providers have a set of standard discounts offered to any customer meeting the discount conditions. Seasonal or special offer discounts can be run in conjunction with promotions.

Promotions are usually offered for only a short time and are intended to accomplish short-term goals. They may also be used as a quick competitive response. You can use promotions to encourage customers to migrate from one voice service to another or to attract new customers to your service. Some possibilities for promotions include:

- Giveaway or discounted CPE equipment
- Installation and/or service activation provided free
- One month of free voice service
- Free or discounted data service for a month
- Free service upgrade
- First three calling features free (or included with service)

You may offer special promotions only to certain types of customers (e.g., those with two to eight business lines) to help establish a presence in a new market. The general rule of thumb is that if you run a promotion for longer than three months, it should probably become a standard service feature.

Finally, this chapter would not be complete without including an industry analyst's opinion. TeleChoice recommends that service providers follow the methodology shown in Figure 8.2 for creating a sound marketing communications strategy.

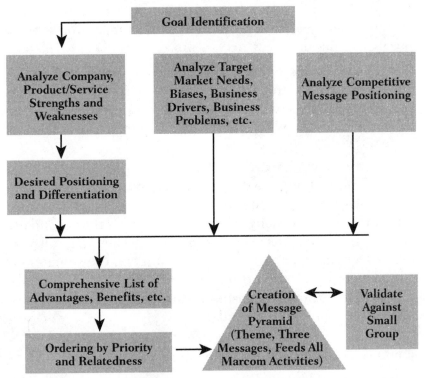

FIGURE 8.2   Creation methodology for a marketing communications strategy.

THE NEXT GENERATION—NOT BEING LEFT WITH STRANDED CAPITAL

CHAPTER
NINE

WHAT TRULY
NEW SERVICES
AND WHEN?

With technological change happening in Internet time, how can small and emerging service providers protect themselves from being left behind with equipment that is difficult to upgrade and threatens to make obsolete their network and services? Moreover, what new services should young carriers be on the look out for to enhance their existing investments in networks and equipment? This chapter examines next-generation trends and services in telephony, and tries to forecast what lies ahead for telecom companies in the next 12 to 18 months. I have included, where appropriate, the names of some telecom equipment vendors that are preparing to introduce equipment for next generation "all-IP" networks and a brief description of new IP telephony services that are on the horizon. This list is by no means exhaustive or all-inclusive, but simply random examples to illustrate certain trends taking place in the arena of enhanced IP services.

In trying to determine the forces that shape emerging services and technologies, some basic and fundamental issues pertaining to the use of bandwidth are explored next, starting with the availability of fiber for deploying new services.

IS FIBER READILY AVAILABLE TO DEPLOY SERVICES TODAY?

The short answer is no. Between 1996 and 2000, ambitious broadband transport companies like Level 3, Qwest, Global Crossing, Metromedia Fiber Network, and others, spent nearly $100 billion to install 39 million miles of fiber in the U.S. and another 60 million miles worldwide. This seems to be a phenomenal amount of fiber—does this signify availability of fiber bandwidth for new services? Not really. First, bandwidth refers to usable or provisioned pipes. What exists today is mostly unprovisioned dark fiber in the ground. Ninety-five percent of this unprovisioned fiber is going unused. Until this fiber is provisioned or made usable, it remains unavailable for carrying new services. Second, if this unlit fiber is not used soon, most of it is likely to become obsolete before it can be used by the new transport technologies and network equipment that are continually being introduced into the marketplace. At any rate, provisioning this unlit fiber is going to demand substantial spending outlays by carriers—spending that is largely dependent on the financial health of carriers, which in turn depends on customer demand, which in turn depends on a booming economy. Although demand continues to increase, particularly in the business sector, the environment over the past two years has certainly not been conducive to the rapid adoption of fiber rates. The financial burdens of making fiber available to the masses, coupled with the lack of usable applications that would justify its use, and rising telco bankruptcies, have all contributed to fiber's remaining unprovisioned and unavailable. Thus, new services are being deployed, to a large extent, using existing alternatives such as DSL, wireless, cable, and T1.

In conclusion, from an end-user perspective, the vast majority of businesses do not have access to high-bandwidth services today, in spite of the vast amount of fiber that has been installed in the ground in anticipation of demand. But there is the future promise of freely available fiber services, low prices, blazing-speed applications, and increased choices. From a transport carrier perspective, however, there is limited upside.

The pipes are already in the ground and the transport companies that have laid the pipes are in a quandary because of the cap on new cable installation; they are looking elsewhere for expansion and differentiation. Companies such as Qwest and MCI Worldcom have done this by focusing on the services aspect. Those who have not diversified in time risk facing a now familiar plight: plunging revenues followed by bankruptcy.

IS THERE A DISRUPTIVE ACCESS TECHNOLOGY ON THE HORIZON TO DELIVER KILLER APPS?

Yes. It's Gigabit Ethernet (GigE). GigE is an emerging disruptive access technology that changes the definition of speed. There are no incremental improvements with GigE, but a new dimension of performance. Although GigE, like any IP technology, faces QoS and reliability issues at present, GigE brings powerful advantages to the table: a single protocol requiring fewer network elements; low-cost, easier provisioning than DSL or T1; and highly granular bandwidth provisioning, which makes it a lucrative medium for offering multiple high-bandwidth services over the same pipe.

Who will win the GigE competition? Will it be held hostage to the same issues that plagued DSL? It appears not. For one, GigE has changed the rules of the game in the local loop. In the area of GigE, which uses dark fiber, the competitive carriers that survive may now have a distinct advantage over the ILECs. This was not the case in DSL, because the ILEC was in the driver's seat both from a supplier of core infrastructure standpoint and from a "years of technology knowledge" standpoint. The ILECs control the copper infrastructure and were well equipped to effectively thwart the CLECs' attempts to enter and gain market share with DSL-based services. GigE, however, is a new kind of network that has, for now anyway, leveled the playing field for incumbents and competitors. Neither the incumbents nor the challengers have a fully developed core competency in GigE or optical Ethernet that can be quickly leveraged. Frustrating to the ILECs is that, unlike DSL, GigE does not use the existing copper local loop that the ILECs control and understand so well. For building the GigE infrastructure and related core competencies and acquiring the necessary technology "know-how," the CLECs (or ELECs as they are called) appear to have an edge because they are leaner, swifter, entrepreneurial in spirit, and (moreover) do not have a legacy base to protect or contend with. No cannibalization issues exist for the CLECs. They can move quickly, build partnerships, and supply services to metro pockets where bandwidth demand is thriving. Finally, the CLECs do not have to undergo the same debilitating facilities arbitrage with the ILECs to get into the GigE business; this arbitrage proved to be a huge obstacle in DSL deployment. GigE is the battlefield where the war for enhanced services market share will be fought, and it may well prove to be the platform from which the CLECs emerge triumphant. Some notable CLECs (ELECs) that are in the spotlight with early GigE offerings are Yipes, Telseon, and Cogent Communications.

GigE's Killer App Opportunity

GigE will enable the next generation of bandwidth-hungry services, which rely on speed and standards to deliver their full benefits. Customized Internet access, content distribution, enhanced storage services, streaming media, online multimedia presentations, and VoIP are only some of the commercial applications that small and medium businesses will embrace. Fueled by demand for these applications, the size of the GigE opportunity appears immense, growing exponentially in the foreseeable future from $4 billion last year to an estimated $25 billion by 2005. If VoIP becomes the voice of the future, as it is widely believed it will, then VoIP may well be the killer application that drives the demand for GigE. Analysts predict that the VoIP market may be as large as $30 billion over the next few years.

ARE THERE ANY TRULY NEW APPLICATIONS ON THE HORIZON?

Yes, fueled by the open protocols of softswitches and the bandwidth of GigE.

When I asked McLeodUSA what they believed were going to be truly new applications, McLeod had this to say:

> "I wish I had a crystal ball to answer this question! There are many independent events happening today that are driving the adoption of converged voice and data services. These include:
>
> - Voice over IP (VoIP): More and more companies are building internal (private) IP networks to carry their own voice traffic.
> - Large corporations have been deploying TDM-based networks that carry both voice and data services for a long time (e.g., a dedicated DS-3 access pipe that is split by equipment on both ends into voice and data components).
> - DSL is gaining acceptance in the residential and small business markets."

Voice and data convergence are finally due to arrive at the consumer and small business doorstep. The first form of convergence was enabled through Voice over DSL (VoDSL) or Voice over Broadband (VoB) gateways that enable packet voice in the last mile. VoB provided an early and important step in the move to the true convergence of voice and data. VoB does not provide lifeline support, but that is not a big issue for businesses that can afford to have battery backup or a POTS phone sitting in the closet. VoB mimics Class 5 functionality completely transparently to the end-user, with a high degree of reliability and voice quality; when coupled with additional phone lines, a lower price/subscriber line, and a single bill for voice and data, this can be a big plus for subscribers. However, VoB is a last-mile solution based on the GR-303 approach and was architected to facilitate the migration from circuit-switched to packet networks. It is thus a viable solution only as long as

Class 5 switches and analog phones remain in existence. Because of the transparency of its approach, VoB does not support the new generation of IP phones and enhanced services, such as unified messaging and VoIP. These services are best enabled by the new switches.

To leverage the cost savings associated with the use of the public network, and to enjoy low termination rates by bypassing the traditional circuit-switched network and associated toll charges, voice must travel packetized over the entire network, not just in the last mile. Softswitches are already playing an important role in enabling voice and data convergence. However, softswitches are still in very early stages of development and testing, and in fact most softswitches have still to see a full-scale live deployment. The next section discusses where softswitches are today, and which new applications are likely to emerge as winners in the near future.

SOFTSWITCHES: READY TO BE CLASS 5 SWITCHES?

While the telecom network readies itself for the next generation, and this process will take considerably longer than it did in the computer industry, several interconnection components such as media gateways and broadband gateways have become invaluable in bridging the existing circuit-switched network with the new packet-switched network. In the longer term, these components will cease to exist as their functionality gets absorbed into softswitches. Although softswitches serve more of a bridging function today, rather than operating as Class 5 switches, as the transition from the traditional TDM network to an all-IP network moves forward, softswitches will steadily replace the legacy switches in the central office. Today, however, softswitches serve more as "super" media gateways that reside in the central office and facilitate the transition to next-generation packet networks by migrating circuit-switched voice networks to data networks that support packet voice. Softswitches can, and do, interface to the SS7 network, providing a valuable Class 4 trunking function; however most softswitches today do not have complete Class 5 features and thus are not ready to replace legacy Class 5 switches at this point. In the meantime, softswitches function well as media gateways and are being used by both ILECs and facilities-based CLECs to perform useful translation and routing functions such as Internet offload. In this application, instead of terminating on the Class 5 switch, dial-up modem traffic is diverted or offloaded by a softswitch sitting in front of the ingress Class 5 to the ISP, thus freeing up the ports on the Class 5 switch for POTS traffic. Several softswitches such as Santera, Coppercom, Convergent Networks, Rapid5, and Taqua provide this functionality today. Class 4 trunking is another popular application for softswitches. Both Internet offload and Class 4 switching are "migration" applications that facilitate the transition from the legacy POTS network to the next-generation all-IP network by supporting both circuit-switched and IP traffic.

When operating as media gateways, softswitches can provide useful features such as digit collection; dial tone; inspecting and routing voice calls as SS7 IMTs to the PSTN; routing data traffic to the Internet; and offering support for some Class 5 features such as caller ID, call waiting, call forwarding, call blocking, three-way calling, call transfer, and so on. However, a softswitch requires IP equipment at the customer premises, such as an IP-PBX or an IAD, or an IP phone. For this reason, PBXs with hardwired T1 CAS interfaces, coin payphones, or ISDN phones cannot interface to softswitches and require connections to legacy switches.

Without a doubt, the revenue-generating services of tomorrow will be based on softswitches. Today, the business case for softswitch deployment depends primarily on how visionary and IP-based your service strategy is, and whether you can build on the somewhat premature softswitch applications available today to deliver a full-fledged service offering as the technology matures. Conversely, you may want to stay away until the application demand has been validated by customers and the technology tested for reliability, support, and quality. Regardless, as you move to enhanced services softswitches will play an important role in your service strategy. Table 9.1 lists some softswitch-enabled applications.

TABLE 9.1 Softswitch-enabled Applications

CUSTOMER APPLICATION AND USE	APPLICATION FEATURES	SOFTSWITCH VENDOR SUPPORTING THE APPLICATION
Internet offload: A media gateway or softswitch in front of a Class 5 switch, inspects and offloads/diverts Internet traffic to the ISP network , freeing up ports on the Class 5switch for POTS traffic.	This application is transparent to the end-user, and deployed solely for the benefit of the service provider.	Most softswitch vendors support this application.
Class 4 trunking/packet tandem switching: The softswitch in front of the Class 5 switch works in a migration role, diverting SS7 traffic to the Class 4 switch before it can terminate on the Class 5 switch.	This application is transparent to the end-user. It is deployed solely for the benefit of the IXC or long distance provider, to avoid ILEC trunk access and termination charges.	Most softswitch vendors support this application.
IP-PBX: PBX or Centrex that functions both in circuit switched and IP telephony networks, and with IP and POTS phones. Resides at customer premise or in building basement. Can also be offered as a remote hosted IP-PBX offering to small/medium business customers. Devices required at Customer Premise: IP-PBX Traditional POTS or IP phones	Complete business phone functions of traditional PBXs: • DID, short digit dialing, multiple lines, and call appearances • Caller ID, call hold, call transfer, call waiting, call forwarding, park, group pick-up, do not disturb, hold, mute • Call back queuing, speed dialing, three-way conferencing And new features made possible by IP such as: • Unified messaging • Find me/follow me services • Presence services	Sylantro, Broadsoft, Cisco, Vocal Data, NetCentrex, NCI, and others

CUSTOMER APPLICATION AND USE	APPLICATION FEATURES	SOFTSWITCH VENDOR SUPPORTING THE APPLICATION
VoIP: Long distance packet voice between two customer end-points (international or state to state); commonly uses the public Internet backbone network, but can also use the carrier's private backbone network (VoVPN) for greater reliability and QoS. Offered as a hosted service by long distance carriers. Devices required at customer premise: IP or traditional phones IP-PBX IP IAD (for traditional phones)	• Web-based service provisioning and management • Web portal with productivity tools for front-desk personnel • Long distance and international phone calls that use the Internet or IP network rather than traditional PSTN/SS7 network to reach their destination • In the absence of their own IP telephony facilities, carriers can now wholesale VoIP services from different affiliate carriers to increase global call coverage and presence.	iBasis (wholesales to providers), Cisco, VocalTec, NetCentrex, NCI, BayPackets, and others
VoIP over VPN (VoVPN): Long distance packet voice over carrier's private IP network; the IP network shares data services such as VPN. Devices required at customer premise: IP or traditional phones Legacy or IP PBX IP IAD (for traditional phones)	Features are the same as VoIP above except: • Carriers can offer upgraded SLAs and QoS guarantees since reliability issues associated with public network are avoided. • Businesses can take advantage of the shared private network to tie together voice services with their VPN services in the different branches, subsidiaries, and headquarters.	VocalTec, Cisco, NetCentrex, BayPackets, Broadsoft, and others

continued on next page

TABLE 9.1 Softswitch-enabled Applications (Continued)

CUSTOMER APPLICATION AND USE	APPLICATION FEATURES	SOFTSWITCH VENDOR SUPPORTING THE APPLICATION
Unified Messaging: Messaging systems (fax, voice, e-mail) are integrated in a single communications environment accessible from phone, browser, or e-mail client. Devices required at customer premise: IP or traditional phones IP or legacy PBX IP IAD (for traditional phones and legacy PBX) PC with Web browser	• Access voice messages via phone or e-mail • Viewing, managing, and accessing of voice mail from phone and Web interface/e-mail • Send faxes via e-mail • Notification for voice mail by multiple devices such as cell phone, pager, e-mail, etc. • Web-based provisioning of subscribers and mailboxes • Security (secure authenticated access)	Vocal Data, BayPackets, iBasis (wholesales to providers), Cisco, and others
IP-Based Call Center: Hosted virtual Call Centers that provide comprehensive customer support management functions for businesses via advanced GUIs, and virtual help desk functions such as interactive voice response, automatic call distribution, CTI, etc., and can be deployed anywhere in the operator's network. Devices required at customer premise: None additional.	Virtual Call Center with multimedia capabilities: • Advanced graphical user interface • Interactive voice response • Automatic call distribution • Unified messaging functions • Phone communications from and to the Call Center agent use the VoIP network.	NetCentrex, VocalData, Cisco, and others

Customer Application and Use	Application Features	Softswitch Vendor Supporting the Application
Hosted Multimedia Conferencing Services: Carrier-hosted conferencing services for small and medium businesses. Devices required at customer premise: Conferencing equipment for viewing. IP or traditional phones	• On-demand video/audio conferencing • n-way conferencing	Vocal Data and others
Pre- and post-paid VoIP calling cards: Calling cards based on VoIP rather than traditional long distance voice. Devices required at customer premise: IP or traditional phones IP-PBX IP IAD (for traditional phones)	The calling card works just like traditional calling cards, except rates are lower. Service providers benefit by bypassing the circuit-switched network and paying low termination rates, and potentially reaching a broader customer base.	VocalTec and others

SOFTSWITCH APPLICATION MANAGEMENT TOOLS

- *Application Programming Interface (API) for service creation.* Softswitch provides a customizable API for modifying existing services for customers and creating new services from scratch using pre-built service templates.
- *Provisioning and management.* Softswitch provides a Web interface or GUI (graphical user interface) for self-provisioning by customers for near real-time control over service features and simple management tasks such as the ability to add or delete users.
- *Security.* Softswitch provides the capability to authenticate user access to services via a username and password.
- *QoS management.* Softswitch provides capabilities to manage and/or monitor QoS implementations in the network such as DiffServ, RSVP, COPS, and MPLS.

GIG-E ENABLED APPLICATIONS

Figure 9.1 shows the continuous rise of bandwidth-hungry applications. While it may appear that most bandwidth intensive applications, such as video on demand and interactive shopping, are likely to be more desirable to residential consumers, small and medium businesses are a lucrative target segment for high-revenue value-added applications such as VPNs, Web hosting, storage services, VoIP, video conferencing, and telecommuting at the speed of light. Interactive shopping and video on demand are applications that are increasingly desirable for both residential and business customers. The high-bandwidth transport medium also opens up all kinds of possibilities for ASP. ASPs can start offering the same level of service as internal corporate computing centers.

ECONOMIC AND REGULATORY CHANGES

The current economic environment is changing how large enterprises think about purchases. More than ever, these cus-

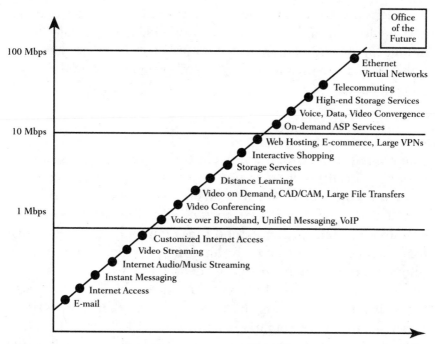

FIGURE 9.1 Progressively bandwidth intensive services.

tomers want the biggest bang for their buck, and the cash crunch is forcing them to look at creative ways to fund their new purchases. With increasing capital and IT budget constraints, more and more companies are interested in outsourcing IT infrastructure and services instead of building these capabilities themselves. Carriers that focus on providing and maintaining these applications will most certainly thrive in these tight economic times. Application service providers, managed services providers, and Web hosting and e-commerce providers are some of the types of providers that should do well in the coming years.

Small-business customers are also changing their purchase habits. Because of the economic slowdown and the failure of multiple carriers and equipment makers in the telecom space, customers are wary of farming their telecommunications business to a service provider or equipment maker who may not be there to support their service in the coming months. Service providers must therefore be prepared to demonstrate financial and business viability to their customers. Cutting-edge technology should be demonstrated to prospective customers only to show cost savings and operational efficiencies, and not for its own sake.

REGULATORY CHANGES

The regulatory environment is almost certain to change in the coming year although it is not clear whether the changes will favor the competitive carriers or not. That said, it would be a mistake to allow the ILECs to remonopolize the local phone market and then further extend that monopoly to the long distance market before a truly level playing field in the local phone markets has been established. The CLECs as a collective body do not have the powers to make this happen. Thus, the burden lies on regulators to ensure that competition increases in the local markets before the ILECs are allowed to enter the long distance market. The ILECs must be persuaded (whether financially, structurally, or legally) to open up the local markets to fair competition. Several suggestions on the type and extent of regulatory participation have been bandied around, including some in the early chapters of this book, but exactly how and when this will happen remains to be seen. The advent of fiber in the last mile and a lesser reliance on the copper local loop is also likely to influence regulatory changes.

Out of several telecom related bills pending a decision, there are currently three significant bills in Congress aimed at changing the rules of the Telecommunications Act of 1996:

- Tauzin–Dingell House of Representatives Bill HR-1542
- Cannon–Conyers House of Representatives Bill HR-2120 (Amendment)
- Hollings Senate Bill S-1364

HR-1542.   The Tauzin–Dingell bill, also called *The Internet Freedom and Broadband Deployment Act,* clearly favors the baby Bells. The bill passed on February 28, 2002.

Considerably weakened by amendment bills such as HR-2120, which would negate its far-reaching effects, the controversial bill proposes to add long distance, high-speed data connections to the list of services regional Bell companies may offer without prior approval from the FCC. In essence, the bill would allow incumbent local exchange carriers to enter the broadband market, extending the Bell monopolistic practices to IP communications. The bill would also eliminate the requirement for Bell companies to open up access to broadband-enabled remote terminals to competitors. The stated goal of HR-1542 is to "deregulate the Internet and high-speed data services, and for other purposes," as well as mandating a 100 percent build-out of broadband networks by the RBOCs within five years. Pointing out how DSL deployment has stalled, the authors of the bill assert their beliefs that this widespread ILEC buildout will bring broadband to the masses and inject billions of dollars into the sagging economy. The bill's authors seem to have overlooked the fact that consumers have ready access to cable broadband where DSL is not available. Also, not all Bells view the 100 percent build-out favorably, particularly those like Qwest, which operates primarily in large rural areas. See Table 9.2.

TABLE 9.2   HR-1542 Opponents and Proponents

TAUZIN–DINGELL BILL HR-1542	
FOR	AGAINST
RBOCs	CLECs
U.S. Telecommunications Association	IXCs/ISPs
Billy Tauzin (R-LA)	Association of Local Telecommunications Service
Equipment vendors	Equipment vendors

Source: RHK

HR-2120.   Bill HR-1542 now contains an amendment that would give the Department of Justice authority to determine whether incumbent carriers should be allowed to offer long-distance data service. The Justice Department so far has been able to make recommendations, but not offer determining decisions. The amendment also clarifies that nothing in the Telecom Act changes the fact that baby Bells are subject to antitrust laws.

Under HR-2120, also called the *Broadband Anti-trust Restoration and Reform Act*, a Bell company engaged in anti-competitive conduct is prevented from providing long distance services. This amendment was definitely a victory for the CLECs, although neither side is as yet confident of the outcome.

S-1364.   Definitely in favor of the CLECs, this bill seeks to improve and strengthen the provisions of the 1996 Telecommunications Act. The bill was predicated on the following findings by the FCC:

· Six years after the 1996 Telecommunications Act came into being, the Bell operating companies have met the local phone market opening requirements of that Act in only five states.

· It is apparent that the Bell companies do not have adequate incentives to cooperate in this process, and the regulators have so far not exercised their enforcement authority to require compliance.

· With improved mandatory penalties on Bell companies and their affiliates that have not opened their networks to competition, there will be greater assurance that local phone markets will be opened more expeditiously and, as a result, consumers will reap the full benefits of competition.

The goals of the Bill are to improve and strengthen the enforcement of the Telecommunications Act of 1996, to ensure that local telecommunications markets are opened more rapidly to full, robust, and sustainable competition; and to provide an alternative dispute resolution process for expeditious resolution of disputes concerning interconnection agreements.

Under this bill, the FCC can punish those Bell companies found in violation of the 1996 Telecommunications Act. The

FCC can now impose an increased penalty of $10 million (from the current $1.2 million) for each such violation, and an additional $2 million for each day on which the violation continues after issuance of the order. Repeat offenders would receive triple penalties.

As another major remedial measure, the bill also advocates splitting the Bells into wholesale and retail divisions, with separate accounting and business transactions for each division.

WHERE TELECOMMUNICATIONS IS HEADED

The telecom industry is a decade behind the computer industry in opening up proprietary telecom networks to standards, distributed models, and interoperability. We are just beginning to see in telecom what happened in the computer industry many years ago; just as the monolithic proprietary mainframes gave way to standardized client–server distributed systems, the monolithic Class 5 switches are slowly giving way to softswitches that resemble the distributed client–server computing architecture in many ways.

Softswitches use open standards, such as SIP and MGCP, to communicate with a variety of distributed best-of-breed platforms that enable a new generation of IP-based services and reduce the cost of delivering existing phone services. These new voice applications cannot run on today's Class 5 switches, thus mandating a switch from the old to the new. Ahead of the switching model change, the service delivery model has already changed from granting control to a single manufacturing entity (the telecom equipment manufacturer) for both switching platform and services to a number of "specialized" software and hardware entities that compete for the development and delivery of services and switching platforms to service providers. This has broken the stranglehold on service providers that the switch manufacturers had for decades, and ushered in a new era of innovation and competitiveness that has resulted in lower prices and better service availability. The beneficiary of these changes is both the service provider and the end-user, but these changes have also enabled, for the first time, a new breed of telecom-focused ISVs that are developing applications at a fast and furious pace for next-generation open switches. This new service delivery model is, in part, the driving force behind the switching model change from monolithic switching platforms to open, distributed platforms.

Technologically, we are in the midst of the most disruptive change that we have seen since the Internet. There is a major shift occurring from the old copper-based infrastructure to a new optical IP-based infrastructure that will be used to carry broadband converged voice and data traffic. Much to the benefit of the younger competitive carriers, the existing central office, along with the legacy Class 5 switch, is on the way out. This sea change will be collectively driven by the new optical infrastructure that will bypass the legacy central office switches, and by changes in the service delivery model described above.

CHANGES IN TELECOM SERVICE CREATION

The telecom service creation environment is finally opening up. We have come a long way from being forced to adhere to Bellcore SCE specs for every service we wanted to develop, test, and deliver. Services can now run in a shared state on multiple devices including the softswitch, PC, cell phone, PBX, CPE appliance, and so on. Most new services are written around flexible and open protocols, such as IP and SIP, and the emphasis is on interoperability in a distributed peer-to-peer network. Some of the key changes that are taking place in the service creation infrastructure include:

- *Provisioning.* In today's fast-paced "on-demand" environment, customers are no longer willing to wait for days to get their new service going. This is particularly true for the "pay-per-usage" type of applications such as bandwidth, storage, conferencing, and so on. The newer services being rolled out are thus geared to support this type of *automated real-time provisioning* that will minimize, if not eliminate, the costly time lag between service order and service turn-up.

Another feature being built into softswitch services is the ability to *self-provision*. More and more, even as businesses outsource their service needs to providers, they want to retain the ability to manage certain components of the service, such as adding or deleting users, adding or deleting service components, viewing their bills online, and so on. New softswitch services typically include a graphical user interface (GUI) or Web-based provisioning interface or portal that will allow end-users to self-provision any of the Class 5 service features or select and order enhanced service features via a Web interface, thus granting them instant assignment, availability, and control over their services. Through APIs in the softswitch, service providers can customize these self-provisioning features to meet customer needs.

- *Metering.* The new services will require a new approach to billing. To extract compensation for value in an intensely competitive scenario, service providers will need to consider alternative billing methods such as charging by packet, byte, bandwidth, or application usage. The new converged services bills will most likely include combinations of different billing methods for voice, video, and data. The provisioning and management software provided by infrastructure and switch equipment vendors will need to include capabilities for metering usage of applications and packets transmitted as well as providing true flow-through provisioning capabilities to manage SLA parameters.

- *Customer profiling.* Some softswitches already support multiadministrative domains (i.e., partitioning) to allow mass customization and LDAP data store of customer profile information.

- *Zero-touch order processing.* A new way of ordering services will allow customers to order their services on the Web. Instead of a customer service rep taking the phone call and then following a manual process for verifying the order and activating the provisioning process, the orders are automatically entered into an automated back-end provisioning system that completes the order processing and provisioning process without requiring manual intervention.

WHAT DO CUSTOMERS WANT?

What do small businesses need by way of access technology? Small businesses have hailed the advent of DSL, although DSL has failed to deliver in many instances (DSL problems and potential solutions are described in the next section). Small businesses are cost sensitive, and the smaller the business the more influence price has over reliability. Where available, and if businesses can tolerate the slow provisioning times and reliability issues associated with it, DSL's average cost of $100 to $200 per month is a very attractive alternative to a T1 line that essentially delivers the same bandwidth (1.54 Mbps) at about double the cost, although it comes with guarantees of higher reliability. While T1 is fast catching up in price, and you may be able to wholesale T1 circuits for much less than was previously possible, DSL is still cheaper and will, in most instances, deliver the same bandwidth for half the price (see Figure 9.2).

Moreover, with the advent of VDSL (very high bit rate DSL), which supports blazing download speeds of up to 26 Mbps, video and broadcasting applications are becoming a more exciting possibility. VDSL offers bandwidth plentiful enough so that every customer requirement—multimedia, voice, and data—can be delivered using a single access source. With bandwidth speeds ranging from 13 Mbps to 26 Mbps, a carrier can offer multimedia applications, high-speed Internet access, and traditional telephone services simultaneously over a single VDSL connection.

Small businesses in need of the reliability of T1 but unable to justify the expense of a full T1 can lease a fractional T1, or obtain "burstable" T1, which lets them increase bandwidth temporarily in response to extraordinary data flow while still paying a rate based on their average traffic. Combining voice and data on a fractional T1 can cost less than $1,000 a month, a very attractive package for most small businesses. Allegiance Telecom is offering such a service to small businesses.

If you are targeting the medium-sized or large business customer, then you need to know that DSL may not be the appropriate connection technology for them. Larger enterprises have

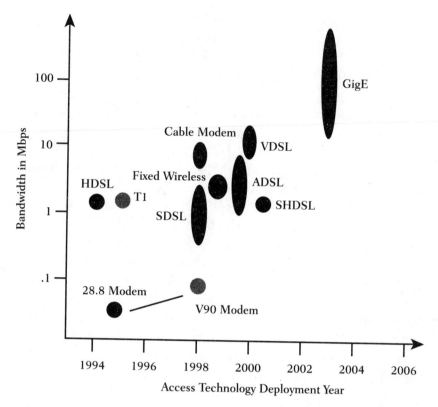

FIGURE 9.2   Access technologies: today and tomorrow.

endorsed the dependability of T1 by staying with it, and while it costs more, it is the preferred choice for companies that depend on constant data links. In fact, a single T1 line with 1.54 Mbps can adequately handle the data traffic of a company with a modestly sized Web site and as many as a couple of hundred e-mail and Web users. The issues with T1 that large customers don't often like are that T1 installation can take many weeks, if not months, and T1 is difficult to expand.

THE FUTURE OF DSL

Last year, providers embraced DSL because it could support multiple services over the same copper wire. It was not only about speed, but an opportunity to introduce new services to the small-business community. With DSL, users could surf the Web faster, download files in much less time, and enjoy new applications like online backups or streaming video. In fact, DSL-based service deployment was perceived to be a win–win opportunity, both for the service provider who could provision value-added broadband services and for users of these services, who would realize dramatic cost and performance benefits.

Service providers embraced DSL for another reason as well. DSL is capable of carrying packet-based voice along with data, and this offered a unique opportunity to new and competitive service providers to capitalize on the $46 billion LEC bypass marketplace. This new DSL technology, Voice over DSL (VoDSL) threatened to revolutionize the business telecommunications marketplace last year. After all, VoDSL would allow you to offer small- and medium-sized businesses high-speed data and up to 24 voice channels over a single copper telephone line. And VoDSL was the first technology that promised the true convergence of voice and data in the last mile. The sad truth is that despite its promise, DSL floundered and did not take off as expected.

DSL HURDLES AND HOW TO OVERCOME THEM

DSL deployment has suffered some major deployment setbacks over the past couple of years. First, DSL deployment ran into problems early on, delivering neither the reliability, nor the coverage, nor the short turn-up times promised by service providers. DSL connections were inconsistent, with poor support. CLECs, eager to exploit the new broadband opportunity and make billions, rushed to sign up as many customers as possible, far outstripping their ability to provide good service. DSL deployment thus stalled, in part because of foot dragging by the incumbents and in part because of the lack of preparedness on the part of the CLECs. This severely impacted the rollout of the high-speed data and converged services dependent on DSL. When revenues did not materialize as expected, most of the cash-starved CLECs offering DSL went out of business, forcing consumers and businesses to pursue other avenues for broadband access, such as T1, wireless, or cable.

The burning question now is, can providers overcome the obstacles with DSL turn-up and provisioning? There are many opinions. According to TMNG's Marsh, the fact that a truck must roll to put in a DSL connection is the main limiting factor here: the real failure of the CLECs and ILECs was that they didn't accurately measure distances from the COs and have accurate maps of where DSL can be supported. By preparing this type of information, CLECs could have overcome some of the DSL provisioning hurdles. "If DSL cannot break out of its limitation of distance from the CO, it will be of limited use," observes Marsh. "Small businesses that require Internet access will migrate to other options, but in fact, these options have the same issues as DSL, such as time-to-install and provisioning issues. I suspect they will migrate to fiber as it becomes available for all their data needs since fiber is looking to be more dependable and will support growth more easily than DSL. Ultimately, I think DSL will go by the wayside. Fiber will replace standard copper and be able to support broadband services more easily. There will be the same provisioning issues, but I believe the DSL experience will smooth the process for fiber."

McLeodUSA is deploying DSL-based services to the small and medium business. McLeodUSA management believes that DSL has a definite future as an access technology in the small business marketplace. "All technologies have pros and cons, but the ability to integrate voice and Internet access in a cost-effective manner is a big plus for DSL. In addition, many of the service activation hurdles that providers have faced are behind us. Interestingly enough, T1 services can be offered *via* DSL more cost effectively than through traditional methods."

On the subject of DSL provisioning hurdles, McLeodUSA had this to say:

> The key to overcoming the provisioning hurdles associated with DSL is in the deployment of physical facilities from the colocation to the customer premise. Most network technology today will allow carriers to cleanly provision customers from their colocation inward; the key will be to overcome the issues that are posed by utilizing the ILEC's

'last mile'. The facilities issues related to the ILEC are generally associated with one of several factors: lack of copper facilities, conditioning of the pairs, accurate distance from the central office, and continuity of the facility. In order to drive these issues and empower CLECs to become better equipped to provision DSL, I believe CLECs need to take several actions:

- Acquire all available electronic tools from the ILEC— EDI, loop qualification tools, on-line ordering systems, and any LFAC (Local Facility) database information.

- Automate internal systems to accommodate 'flow through' of ordering information. The key for the future will be utilizing technology to reduce cost to make broadband less expensive and more attractive to the masses.
- Continue to drive the ILEC toward greater openness of their networks; this would be accomplished through business-to-business negotiations, PUC and FCC filings, merger, and 271 workshops.

Deployment of fiber will actually do very little to assist in the deployment of DSL nationwide. The main reason is that DSL is primarily dependent on the copper facilities from the ILECs. Although fiber connectivity will assist in transporting traffic between hubs, colocations, and ISPs, it will do very little to help transport traffic from the customer premise to the DSL 'cloud.' The one exception to this position is the concept of opportunistically deploying fiber optic routes near dense population centers such as neighborhoods, and building complexes, and using copper or coaxial connections from the fiber route to the premise. Obviously, this can be a costly venture with no guarantee of return, duplicating the existing ILEC infrastructure. and adding cost to the deployment of DSL.

Will DSL cease to exist as a viable technology, and have an extremely short life span, like ISDN? Clearly, unlike ISDN, DSL is not a technology looking for a market need. Where available, DSL offers both lower costs and plenty of bandwidth to meet most customers' demands. I believe that there are other options, however, that will surpass DSL down the road. Fiber is one such option. Nonetheless, today copper wire is everywhere and DSL remains a viable broadband solution for copper infrastructure.

CABLE—NOT OPEN FOR BUSINESS (YET)

Cable is a great alternative to DSL in the residential market, and fast overtaking it. Easier to provision than DSL and backed by large cable operators, it is fast becoming the medium of choice for residential consumers. Cable modems now offer better dependability and more capacity per dollar than DSL— especially for small office, home office (SOHO) workers and telecommuters working at home. Cable, however, is not a viable medium for businesses because most cable lines still don't pass by business districts.

Cable is a shared medium, which also makes it unsuitable for business traffic. Unlike cable, both DSL and T1 are dedicated to a single user, and thus able to maintain connection speeds—although these are modest compared to cable's 10 Mbps. Cable's drawback is that, being a shared medium, it can experience slow speeds as traffic increases.

FIBER

Next-generation access technology will be based on fiber. The explosion in Internet use and other bandwidth-hungry applications has led to a huge demand for the high bandwidth associated with fiber to the curb. That said, only an estimated 5 to 7 percent of small- and medium-sized businesses are served by high-speed fiber, with most of these businesses residing in commercial buildings. The majority of the small-business customers are still battling for bandwidth over old-fashioned copper-based PSTN and leased T1 lines.

If you are considering leasing dark fiber, here is a comparison of leasing and provisioning costs that you will find interesting. A CLEC deploys DSL-based services by installing

DSLAMs in colocation cages adjacent to the ILEC's switching equipment. The CLEC then hauls traffic from each colocation site to a single regional switching center, commonly at DS-3 (45 Mbps) rates. Today, these DS-3 tariffs actually cost more than leasing dark fiber to backhaul DSL traffic from the DSLAMs to the RSC. Moreover, dark fiber can be provisioned to supply 20 times the speed with GigE services.

Here's the caveat with dark fiber, however: the cost of fully provisioning dark fiber can be relatively high compared to traditional DS-3 circuits. The cost may include expensive optical equipment deployed along the network route, as well as additional networking equipment (e.g., multiplexers) to make the capacity available for service. For this reason, service providers will only provision (or light) fiber incrementally as there is demand for that capacity. All the same, fiber represents the route to profitability via next-generation services, and smart service providers should be placing a high emphasis on getting their networks ready for optical speeds.

Large businesses and commercial customers in high-rise buildings, mostly in metropolitan areas, are also increasingly switching to fiber where it is available, for obvious reasons. In the areas of both service provider infrastructure and service delivery to business customers, there is a wide range of applications that place heavy demand on network capacity. Videoconferencing and online multimedia presentations are two examples of bandwidth-hungry applications. Fiber is the technology that enables these applications. The ideal high-speed connection to the Internet is fast approaching 10 Mbps and above, and this is beyond the capacity of most copper connections. Only fiber can support speeds exceeding tens of megabits per second.

TEN PREDICTIONS FOR 2002–2005

1. *Managed services will be big.* Almost every communications service will be offered on a monthly managed basis including managed telephony, managed networks, managed databases, managed application hosting, managed

security, and so on. Rentals are in and straight purchases are out, particularly for small and medium business, but even for many large enterprises.

2. *The Application Service Provider industry is set for a big revival.* The huge up-front cost of deploying CRM, ERP, and other business-critical applications, combined with the necessity of having these to improve business processes, will force companies to look at renting rather than buying. The advent of GigE will be another factor that will drive the resurgence of ASP applications. ASP services are more likely to be offered by carriers rather than ASPs themselves.

3. *Voice and data will finally converge in the true sense.* End-user demand for IP-based services such as VoIP and unified messaging will be the catalysts to make this happen. Carriers will rush to embrace convergence. The availability of GigE in the last mile will be the dominant disruptive force behind this convergence.

4. *DSL will continue to be strong.* DSL is not destined to die the premature death that ISDN did. In time, as fiber adoption rates increase, DSL will become the poor man's fiber.

5. *3G will have lackluster reception in the marketplace, primarily because of the technology gap phenomenon and the lack of usable applications.* By the time 3G applications mature and users embrace the concept, carriers will be preparing to roll out 4G. Thus 4G will be the real disruptive force behind the demand for blazing speed and ubiquitous mobile Internet services. This is not likely to happen until 2004.

6. *Fixed wireless (MMDS) access will fall by the wayside.* Fiber/GigE and DSL will be the dominant broadband access technologies. Longer term, T1 will be superseded by SDSL and VDSL, both of which will offer a better and faster medium for deploying services on existing copper.

7. *As a result of the recent turmoil in the industry and economy, a few super-CLECs will emerge that will become dominant players and give the ILECs a run for their money.*

Longer term, these super-CLECs may take on the shape and form of ILECs. The IXCs will embrace VoIP, unified messaging, and wireless services to boost revenues and keep long distance customers loyal.

8. *More service providers will adopt the "one-stop-shop" service model, driven by the need to maximize revenues from baskets of services.* Rather than invest in developing capabilities, these one-stop-shop providers will depend on partnerships with affiliate providers to deliver the necessary services to their customer base. The emphasis will shift from having a "best of breed" or "specialized provider" image to a "convenience store" image, with a wider array of products being offered by a single provider. As a result, there will be a blurring of lines between the different types of carriers and their respective telecom segments. We can already see this in numerous examples around us, such as the wireless providers offering Internet access and e-commerce, cable operators offering telephony, building centric providers offering ASP services, and long distance voice operators offering data and Internet access. In this melting pot of a provider environment, differentiation will arise from specialization around the types of customers providers seek, such as consumers, other carriers, enterprises, and so on.

9. *After 2004, the ILECs will likely be structurally split into two legal and separate entities, wholesale and retail.* I'm not sure that this will happen, though, for two reasons. One, because even though this is the logical route to an improved competitive environment, it will be driven by regulators and government intervention rather than by customers or technology—and I have much less faith in the former than I do in the latter. Two, a structural split-up may not even be necessary if the newer optical networks render obsolete the existing local loop and infrastructure, which is at the heart of this regulatory debate, sooner rather than later.

10. *Class 5 switch development has already been capped.* Looking beyond 2004, the legacy central office that sup-

ports the circuit-switched network as we know it today will cease to exist. Some changes, such as the advent of softswitches in the central office, are already starting to happen. The catalysts behind this were described earlier. The network of the future will be the all-IP packet-switched network with softswitches. This complete transformation will take at least seven to ten years to complete.

MARKET
FORECASTS

This chapter marks the end of an interesting project. When all is said and done, we frequently look to industry analysts to gauge their reactions to market trends, validate our own thinking, and feel surer about the direction in which we are headed. To that end, this chapter represents the thinking of leading industry analysts in telecommunications—at least as far as the future of broadband technologies and telecommunications services are concerned.

Notably, when projecting forward, you will find that there are very few number forecasts in this chapter, particularly relating to new technologies such as VoIP and GigE, and in these areas the focus is on qualitative rather than quantitative research. Most analysts now shy away from producing quantitative data projections, and who can blame them? The wildly optimistic forecasts about DSL deployment in the past have not come true, and analysts are now markedly cautious about quantitative forecasting. That said, along with qualitative findings, Gartner Dataquest, RHK, and TeleChoice have all provided their most likely number forecasts on the percentage growth of services and technologies.

This chapter thus explores the trends and evolution in broadband service deployment growth, the transitioning from circuit-switched to packet-switched networks, and the primary drivers behind enhanced IP services.

TELECOMMUNICATIONS MARKETS PERSPECTIVE

In general, analysts agree that the market for circuit-switched voice services has matured. With Class 5 spending frozen, emphasis is now on softswitch development and testing, and the consensus is for data services, packet voice, and mobile services to drive strong telecom service growth over the coming years.

Analysts also agree that the telecom fixed-equipment market (particularly DSL) has declined in 2001 and, impacted by continued capital spending freezes caused by the 2000 over-build by carriers, it will likely continue to decline through 2005. Globally, the overall telecom equipment market, while having declined in 2001 by 4.3 percent, will recover and grow by 6.0 percent in 2002 (source: Gartner Dataquest). In the U.S., equipment related to optical networking (specifically, in metropolitan area networks), Gigabit Ethernet in the last mile, mobile networks (3G, 4G), and packet voice technologies (IVR, VoIP) will experience good growth even during this downturn.

Capital spending has slowed for another reason as well. The demise of so many CLECs and other carriers has lessened, if not ended, the dogfight to sign up new subscribers, thus creating a more relaxed competitive environment for the survivor CLECs and incumbent operators. Less fearful of losing customers to the competition, these surviving companies can now focus on bringing their businesses back to profitability, and many are now happily postponing investments to improve their bottom lines.

However, it is much easier for these companies to delay capital spending expenditures than it is to reduce reliance on a service that is a vital part of its core business—namely voice. Demand for voice will remain solid, as will demand for higher bandwidth to facilitate the ever-increasing demands placed by data services such as Web hosting, VPNs, and Web-based productivity applications.

In this difficult environment, the pendulum of influence will swing from the vendor towards the end-user, righting the disequilibria that have existed in the recent past. Instead of vendors pushing underdeveloped and costly new technologies onto service providers and influencing their deployment decisions, service providers will tend to first weigh end-user demand and the adoption rate of a new technology before implementing a new service or technology. End-users will once again find themselves in the driver's seat: End-user tells service provider → service provider tells vendor → vendor builds service platform → service provider deploys → end-user conveys satisfaction → and the cycle begins all over again. Consequently, we are also likely to see more and more applications defined by users, and this will drive the development of more "customizable" features in applications.

Worldwide, telecommunications revenues will remain fairly strong, although not returning to the exuberant growth widely predicted during the late 1990s. Table 10.1 shows worldwide telecommunications market revenue (in millions of U.S. dollars) by region and sector from 1999 through 2005.

THE STATE OF BROADBAND DEPLOYMENT

DSL DEPLOYMENT SCENARIO

For the DSL sector, there is good news and bad. First, the good news: service providers will see their DSL service revenues increase as they sign up new subscribers. There is also going to be less competition in the business DSL space because of the demise of so many CLECs that targeted this space. This will make it easier for the surviving CLECs to sign up subscribers. DSL subscriber growth will continue at a fairly robust rate as small businesses switch from dial-up, ISDN, and T1 connections to SDSL and SHDSL.

Now, the bad news: DSL vendors will not participate in this growth. Because of capacity buildup and continuing spending caps, coupled with declining price/port ratios, DSL equipment vendors will see dramatic declines in DSL equipment revenues.

TABLE 10.1   Worldwide Telecommunications Market Revenue (1999–2005)

	1999	2000	2001	2002	2003	2004	2005	CAGR 2000–2005
Asia/Pacific and Japan	$257,384	$307,272	$344,879	$378,146	$410,609	$446,070	$489,163	9.7%
Central and Eastern Europe	$37,499	$42,666	$49,292	$56,879	$63,883	$67,861	$71,479	10.9%
Latin America	$99,429	$116,097	$129,724	$143,220	$161,641	$184,680	$201,667	11.7%
Middle East and Africa	$51,283	$57,646	$64,300	$72,538	$81,659	$93,256	$108,869	13.6%
North America	$377,404	$432,844	$456,763	$497,401	$545,000	$596,490	$650,623	8.5%
Western Europe	$292,805	$316,647	$328,956	$357,231	$378,843	$401,074	$426,304	6.1%
Total Telecom Equipment	$311,302	$381,030	$364,777	$386,482	$410,650	$443,920	$491,770	5.2%
Total Telecom Services	$804,502	$892,142	$1,009,137	$1,118,933	$1,230,985	$1,345,512	$1,456,336	10.3%
Total Telecom Market	$1,115,804	$1,273,172	$1,373,914	$1,505,415	$1,641,635	$1,789,432	$1,948,106	8.9%

Source: Gartner Dataquest (November 2001)

Telecom analyst RHK estimates that DSL equipment revenues will decline from $1.9 billion in 2000 to about $804 million in 2001 and then further dip to about $125 million in 2005. Revenues for SDSL equipment will decline from approximately $57 million in 2001 to $1.3 million in 2005. SHDSL equipment, on the other hand, will see modest growth, from $4.7 million in 2001 to approximately $26 million in 2005.

DSL DEPLOYMENT GROWTH

Both TeleChoice and RHK provided research on DSL deployment figures. TeleChoice's estimates are based on conducting DSL deployment tracking surveys, and the numbers shown in Table 10.2 reflect the status at the end of third quarter of 2001. These deployment figures do not include HDSL or HDSL2 deployment figures. Another factor that TeleChoice cited in this calculation was that the business versus residential percentages calculated for the entire market may be misleading, because many home office customers are purchasing residential products instead of business-class products.

TABLE 10.2   TeleChoice 3Q '01 DSL Deployment Summary

SERVICE PROVIDER	3Q '01 LINES IN SERVICE	% RESIDENTIAL	% BUSINESS
ILECs—U.S.	3,254,225	80	20
CLECs—U.S.	539,415	42	58
IXCs—U.S.	28,000	15	85
Total	3,821,640	74	26
Canada	928,612	76	24

According to TeleChoice research, there were 3,821,640 lines in service at the end of 3Q 2001 in the U.S. This number consists of approximately 85 percent ILEC, 14 percent CLEC, and 1 percent for IXC customers. In the U.S. and Canada, there were a total of 4,750,252 lines in service at the end of 2Q 2001.

According to RHK, North American DSL subscriber growth will remain solid throughout the forecast period (2001–2005).

RHK expects that residential and business DSL subscribers will grow from 4.75 million at the end of 2001 to 15.76 million by the end of 2005, a 35 percent compounded annual growth rate (CAGR). Most of this growth will occur in RBOC territories in the U.S. and the two largest DSL service providers in Canada, Telus and Bell Canada. Table 10.3 shows the expected DSL subscriber growth (in millions) by type of service.

TABLE 10.3   North American DSL Subscriber Growth (in millions)

	2001	2002	2003	2004	2005
Residential	4.05	5.70	7.76	10.25	13.36
Business	0.70	1.10	1.60	2.10	2.40
Total	4.75	6.80	9.36	12.35	15.76

Source: RHK

NEW SERVICE DRIVERS FOR DSL

In the business sector, small and medium businesses remain the primary targets for DSL service providers. According to RHK, business-class services that will spur demand for DSL include Internet access, Voice over DSL (VoDSL), e-commerce, distance learning, videoconferencing, and secure VPNs for telecommuters. Business class requirements for telecom services are shown in Table 10.4.

RHK believes that, in addition to Ethernet, cable modems are competing with DSL for broadband in the small and medium business (SMB) (one to ninety-nine employees business) segment. The analyst expects this trend, which began in 2001, to continue as more Ethernet solutions emerge and as MSOs (such as Cox, Charter, AT&T, and Comcast) start moving aggressively into the SMB space.

THE SMALL AND MEDIUM BUSINESS MARKET.   Forrester Research values the small to medium-size business market for integrated communications services at between $60 and $100 billion a year. Only one in five of the estimated 6.5 million

TABLE 10.4   Business-class Requirements (Source: RHK)

END-USER REQUIREMENTS FOR SERVICE PROVIDERS	SERVICE PROVIDER'S REQUIREMENTS FOR VENDORS
Network reliability/availability, "always on" data, and voice connectivity	A carrier-class standards-based piece of equipment that is redundant and interoperable with existing equipment.
Value-added network services: self-provisioning, firewall, secure PPP, DHCP, NAT, VPNs, VoDSL, unified messaging, dynamic bandwidth, ISP selection, etc.	Support for MGCP, H.248 (Megaco), SIP, H.323, dynamic bandwidth allocation, QoS, WFQ, and support for IP services via Diffserv, rate shaping as well as providing customizable access list capabilities from both the customer side or NOC.
A cost-effective one-stop shop for SMB services	Flexible hardware, software, and OSS that supports current voice and data standards and applications. In addition, the platform should be scalable enough to allow future services to be added via remote software upgrade or the addition of a plug-in module.
Investment protection	Reusable network components that can address future requirements and network growth.

businesses in this sector is online now. Table 10.5 shows the percentage of companies in each size category. RHK estimates that 71 percent of all firms have 20 employees or fewer, and 81 percent of all firms have 99 employees or fewer. There were approximately 7.1 million businesses in North America in 2000.

TABLE 10.5   Distribution of North American Firms by Number of Employees for Year 2000

0–4	5–9	10–19	20–99	100–499	500+
48%	14%	9%	10%	5%	14%

Source: RHK

Table 10.6   Average Phone Requirements per SMB Firm

Firm Information	Firm Size Category					
	0–4	5–9	10–19	20–99	100–499	500+
Average number of employees per firm	2	6	12	39	193	3,091
Number of establishments per firm	1	1	1	1.37	3.8	52.6
Number of employees per establishment	2	6	12	28.6	50.7	59
Phone lines per establishment	1	3	6	14	25	29
Typical IAD requirements: port size	4	6	10	16	16	16

(Source: RHK)

Table 10.7   Service Provider Segmentation for DSL Services

Service Provider Type	Customer Focus	Drivers
CLEC and IXC	Small to medium-sized businesses	• Leverage ILEC copper infrastructure • Reduce network costs (UNE) charges • Offer new communications packages for SMB and SOHO segments • Increase average revenue per user • Facilitate migration to packet voice for improved network efficiency • Retain profitable voice revenues (local and LD)
ILEC	Small to medium-sized businesses Residential	• Offer new communications packages for SMB and SOHO segments • Digitize the local loop • Preventative measure against competitive threats from other players (RBOCs, CLECs) • Improve operational efficiencies • Offer "triple play" of voice, video and data services

(Source: RHK)

Fiber Deployment Issues

Analysts agree that the industry will continue to see healthy growth in bandwidth demand, even under very conservative scenarios. Predicting growth in fiber deployment, though, appears not to be as straightforward an exercise as predicting DSL growth rates, and in fact is ephemeral at best. This is because fiber deployment is likely to be impacted by factors other than end-user demand for high-bandwidth services. There is also the widely debated issue of capacity and the high provisioning costs associated with fiber, which make it much more critical that fiber service providers plan their approach more carefully than their DSL or cable brethren.

On the issue of capacity, while there seems to be widespread industry consensus that installed fiber does not equal "lit" or provisioned fiber, identifying those high-growth regions where fiber is likely to be most attractive for lighting depends on several factors, namely availability of installed fiber, corresponding lit transport capacity demand, and network completion reach. First, carriers will be most motivated to light fiber where demand not only exists but outstrips supply by a good measure, and also in areas where somewhat optimal business conditions exist for rolling out lit fiber, such as internal resources (people, cash, floor space), favorable deployment landscape (competition, network reach, partners), and availability of dark fiber. Because the demand for lit transport capacity is ultimately driven by demand for high-end services, fiber service providers must understand capacity in terms of the types of services that will be deployed and the load these services will place on the transport, particularly at peak times. This is one of the key observations behind TeleChoice's Model for Advanced Capital Planning (MADCAP) model for fiber service providers.

The MADCAP model tries to develop a comprehensive view of lit transport capacity and the demand for that capacity, using 22 primary routes connecting 12 major U.S. telecom cities. The objective of the model is to project where deployment growth is most likely to occur in the U.S. by identifying demand and supply variables for lit transport capacity on a route-by-route basis. Based on this, it then calculates the amount of transport capacity demanded across each of the 22 routes in the model. The model also examines 28 different types of bandwidth intensive applications that place a load on backbone networks, and creates a nationwide view of peak application traffic based on these services. For a carrier looking to gauge fiber deployment demand and supply in the U.S., MADCAP claims to determine what the current supply is, and the different applications driving bandwidth on those routes and how fast they are growing. Williams Communications is using the model to help guide capital equipment deployments on its nationwide network.

The 28 applications analyzed by this model are shown in Table 10.8.

TABLE 10.8   MADCAP Model

28 APPLICATIONS INCLUDED IN MADCAP:

Collaborative computing	Network-based PC backup
E-mail	Online file access
Enterprise data networks	Peer-to-peer networking
E-book downloads	Software downloads
Fax calls	Teleconferencing
Film distribution	Teleradiology
Hosted network-based applications	Television backhaul
Image downloads	Television forward distribution
Instant messaging	Video conferencing
Interactive gaming	Video downloads
Long distance voice	Video on demand
Mobile wireless data	Web browsing
Mobile wireless voice long distance	Web casts
Music downloads	Web page posting

Source: TeleChoice, Inc.

Interestingly, the study shows that out of the 28 network applications above, just seven currently represent over 99 percent of all backbone capacity purchased today. These applications, listed in order of current bandwidth levels, are:

· Web browsing
· E-mail
· Enterprise networks

- Long distance voice
- Mobile wireless
- Web casts
- Peer-to-peer

MADCAP CAPACITY DEMAND PROJECTIONS

These projections are based on several specific telecom expansion scenarios (see Table 10.9).

DOOM AND GLOOM GROWTH SCENARIO. This scenario assumes that capital markets remain closed to all telecom infrastructure activities, both long haul and local/metro. The impact of this scenario is that demand growth is constrained by the continued bottleneck in the metro and access portions of the network. Demand continues to grow at a healthy rate, but not at the potentially explosive rates indicated in the other two scenarios. In this scenario, networks (long haul and local) are still upgraded to stay slightly ahead of demand.

RATIONAL EXPANSION GROWTH SCENARIO. This scenario assumes that capital markets are open for local telecom infrastructure activities, but relatively closed for long haul infrastructure. Because of the investment in the last mile, this scenario results in tremendous demand growth. Long haul networks must be aggressively upgraded to keep up with this demand growth; however, long haul carriers stick to success-based growth on their existing routes rather than undertaking large new construction projects. In this scenario, long haul networks are upgraded to stay slightly ahead of demand.

For analyzing capacity growth projections for the 28 applications above, the applications are aggregated into their relevant network "families" (i.e., voice, IP, and other data) to understand the resulting high-level impact of application growth on carrier networks capacity. The network-wide route average summary is provided in Table 10.10 to illustrate overall demand trends in the different scenarios.

TABLE 10.9 Growth Scenarios

BUSINESS ACCESS	2001		2002		2003		2004		2005	
SCENARIO	DOOM & GLOOM	RATIONAL EXPANSION	DOOM & GLOOM	RATIONAL EXPANSION	DOOM & GLOOM	RATIONAL EXPANSION	DOOM & GLOOM	RATIONAL EXPANSION	DOOM & GLOOM	RATIONAL EXPANSION
56.6 Kbps Dialup	33.0%	33.0%	28.0%	36.0%	28.0%	38.0%	29.0%	36.0%	29.0%	35.0%
DSL (256 Kbps)	13.0%	13.0%	20.0%	16.0%	21.9%	19.0%	23.5%	22.0%	24.0%	25.0%
T1 (1.5 Kbps)	13.0%	13.0%	18.0%	14.0%	20.0%	15.0%	22.0%	16.0%	22.0%	15.0%
GigE (1 Gbps)	0.0%	0.0%	0.0%	0.2%	0.0%	1.5%	0.0%	3.0%	0.0%	5.5%

TABLE 10.10 Route Average Summaries

AVERAGE TRANSPORT DEMAND (GBPS) ACROSS 22 ROUTES	2001		2002		2003		2004	2005	
SCENARIO	DOOM & GLOOM	RATIONAL EXPANSION	DOOM & GLOOM	RATIONAL EXPANSION	DOOM & GLOOM	RATIONAL EXPANSION	DOOM & GLOOM	RATIONAL EXPANSION	
Voice	246	246	255	255	265	265	271	271	271
Internet Protocol	851	851	1,291	5,266	1,663	46,607	2,105	138,196	2,548	346,060
Other Data	157	157	218	218	278	278	353	353	447	447
Total Growth	1,254	1,254	1,764	5,739	2,205	47,150	2,729	138,821	3,266	346,778

Based on the findings of the model, TeleChoice believes that, regardless of how quickly the market grows, demand (particularly data) is expected to continue rising at double-digit and possibly triple-digit rates, thereby requiring carriers to continue "lighting" vast amounts of new capacity to meet this demand.

The key drivers behind this application demand growth are:

- *End-user network access speeds (the "metro bottleneck" issue).* This is influenced by the end-user's budget, the speed of the network connection, and the availability of broadband options.
- *End-user adoption rates of applications.* This is influenced by whether or not the application delivers a dramatic benefit for its cost. The cost includes both monetary and end-user (in)convenience factor.
- *Bandwidth consumed per application session.* This refers to the amount and "length" of total bandwidth consumed by an application, which in turn impacts peak loads on networks.

Further, carrier-specific filters must be applied to refine the demand growth variables above, so that carriers can understand not only the market's current and projected revenue potential, but also that portion of the revenue which the carrier is specifically able to address. These filters that a carrier may apply to their own unique situation are:

- (The carrier's) network capabilities
- Service portfolio
- End-user access speeds
- Markets serviced

THE ISP BACKBONE MARKET

Despite their crucial role in carrying traffic to and from the various ISPs, the last 12 months have brought diminishing market capitalization to many of the biggest names in the ISP backbone market, according to Cahners In-Stat/MDR. This high-tech research firm reports that in the face of a forecast zero percent growth in the number of backbone connections used by ISPs in the year 2002, backbone providers will need to focus on the key decision criteria used by ISPs (such as price, reliability, company reputation, company stability, and availability of service) in selecting their backbone providers. The backbone market, along with the total wholesale ISP market, will experience little growth in 2002. The economy is affecting not only the structure of the backbone provider market, but also the services offered and to whom they are offered. Backbone providers are being careful not to repeat the mistakes they made with the free ISP market.

In-Stat/MDR also found that:

· At the end of 2000, ten backbone providers generated 92 percent of all wholesale ISP revenues. WorldCom had the largest market share with 44 percent. The next two largest wholesale ISP service providers were Genuity and Sprint with 12.5 percent and 9.4 percent market share, respectively.

· T-1 remains the most common speed used in backbone connections. This is followed by T-3.

MOBILE BROADBAND GROWTH (WIRELESS LANS, MOBILE NETWORKS)

Although both mobile and Wireless LAN (WLAN) networks provide wireless data access, the similarities end there, according to Forrester Research. This analyst believes that, whereas these networks may serve the same user, that individual will access the two networks with different expectations. Mobile data subscribers will expect connectivity wherever they pull out their handset. Conversely, wireless LAN customers require a worklike setting to do their computing and are willing to seek out the nearest hotspot for access to ensure a quality experience. Their application requirements will also be different. For example, the typical mobile WLAN user is a traveling employee whose work requires broadband access in unpredictable, but well-populated locations—a job for Wi-Fi (see the March 26, 2001 Forrester Brief "Wireless LAN Deployment Gathers Steam "). But when she completes her slide presentation and needs to find a comfortable place to grab a bite, she is much more likely to pull out her handset and click a few buttons to search the local Zagat's restaurant guide.

Devices also will be different for the two types of networks. Forrester predicts that in 2005, 80 percent of the 14.4 million WLAN-accessible devices will still be laptops, dwarfing the number of laptops that use GPRS or CDMA2000 1X modems to connect. Conversely, handsets will dominate access to data via mobile networks (see Tables 10.11 and 10.12).

TABLE 10.11   Mobile Broadband Device Forecast (in Millions of Units)

	2001	2002	2003	2004	2005	2006
WLAN Devices						
Number of laptops	1.94	4.33	6.42	8.67	11.66	16.23
Number of PDAs	0.32	0.79	1.61	2.10	2.79	3.05
Total	2.26	5.12	8.03	10.77	14.45	19.28
Mobile Devices						
Number of laptops	0.60	1.50	2.60	4.00	5.50	7.10
Number of PDAs	0.20	0.30	0.60	0.90	1.60	2.30
Number of handsets	7.70	24.20	56.10	100.10	134.10	187.10
Total	8.50	26.00	59.30	105.00	141.20	196.50

Source: Forrester Research, Inc.

TABLE 10.12   Mobile Broadband Revenue Forecast

	2001	2002	2003	2004	2005	2006
WLAN Revenue						
Total (US $millions)	$40.7	$206.30	$508.90	$953.70	$1,571.60	$2,688.20
ARPU/month	$1.50	$3.36	$5.28	$7.38	$9.06	$11.62
Mobile Revenue						
Total (US $millions)	$0.10	$3.00	$70.60	$648.30	$1,916.60	$3,976.90
ARPU/month	$0.04	$0.07	$0.31	$1.01	$1.80	$2.59

ENHANCED IP SERVICES

Who will participate in the enhanced IP services revolution? According to Cahners In-Stat/MDR—everybody, including the staid RBOCs. Cahners' projections call for the worldwide enhanced IP services market to grow exponentially from $11.1 billion in 2000 to $104.4 billion in 2005 (see Table 10.13).

TABLE 10.13   Enhanced IP Services (WW Revenues in $B)

	2000	2005
Hosting	$5.9	$27.8
VoIP	$1.8	$29.4
VPNs	$1.5	$23.7
Unified Messaging	$0.9	$10.5
Content Delivery	$0.5	$6.8
Other	$0.5	$6.2
TOTAL IP Services	$11.1	$104.4

Source: Cahners In-Stat/MDR

PACKET VOICE OR VoIP

Packet voice is a key component of enhanced IP services. Most analysts believe that the emerging packet voice technology will fundamentally change the way voice is served. Packet voice is increasingly desirable both from an economic standpoint (deploying a softswitch is much cheaper than deploying a Class 5 switch), and because packet voice utilizes open standards and protocols (IP, SIP, MGCP), which makes it very attractive for third-party software and platform developers to build a wide array of new applications. Packet voice, because it behaves just like data, also reduces the costs of maintaining different networks for voice and data.

The single biggest driver of the move to an all-IP network will thus be the economics of serving packet voice (also called VoIP, or Voice over IP). Packet voice will first drive the movement to converged IP networks. Longer term, it will again suffer the effects of commoditization, becoming the "cell-phone giveaway" in a converged services portfolio, as providers throw in voice for free and charge for more lucrative data services on annual contracts.

The IP movement has already begun, and traditional circuit voice providers are suffering the effects of substitute products, commodity pricing, and competition from VoIP as shown in Figure 10.1.

Indeed, when compared to circuit voice, the advantages of packet voice are glaringly obvious, as shown in Table 10.14.

In parallel, research by Forrester shows that new packet voice technologies are showing steady progress, as seen by the following industry developments:

- *Softswitches mimic current infrastructure at a fraction of the cost.* Softswitches from vendors like Sonus Networks and Clarent offer the same switching capability and call-processing

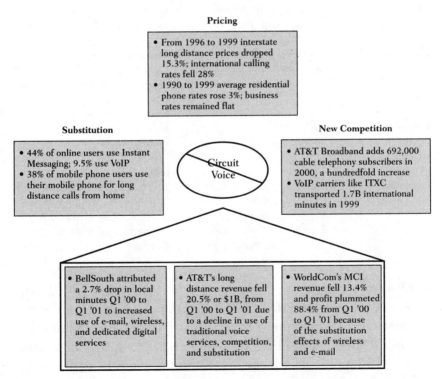

FIGURE 10.1   Effects of substitution, pricing, and competition on circuit voice. (Source: Forrester Research, Inc.)

TABLE 10.14   Packet Voice Upstages Circuit Voice (Source: Forrester Research, Inc.)

	CIRCUIT-SWITCH LIMITATIONS	PACKET-SWITCH ADVANTAGES
Network	• Calls require 64-Kbps connection • Dedicated circuit for each call wastes capacity • Powered for lifeline service	• Compression cuts call bandwidth to 8 Kbps • Multiple calls efficiently share lines for transport • Can be powered for lifeline service
Innovation	• All voice services are identical; differentiation is created by marketing • New apps developed by equipment vendors and bundled with switches	• Independent software developers can introduce applications tailored to individual carriers • Application development cycle can be as fast as several weeks
Standards	• Bell Labs voice communication standards for circuit networks developed mid 1900s • ITU develops telephony standards	• Open standard defined by IETF for application development and call control • IP voice can run on any data network, i.e., frame relay, ATM
Features	• Caller ID, call waiting, voice mail, call forwarding	• Anything that can be encapsulated in IP: voice, video, images, conference calls, unified messaging, "click to talk"
Customer equipment	• Plain old telephones, mobile phones	• Plain old telephones, mobile phones • Any IP device: PC, PDA, call phone, iTV
Network equipment	• Approximately $90 per subscriber line • Proprietary equipment • Highly specialized staff required for maintenance	• $10 to $12 per subscriber line • Off-the-shelf servers can be maintained by standard technical staff network equipment

software as a Class 4 trunk switch. When integrated with servers like Sun's $5,000 Netra hosting call control and applications, they can support many Class 5 switch capabilities, like caller ID and call waiting, for a mere $12 per subscriber, rather than $90 per subscriber. XO Communications and Level 3 Communications already use softswitches to bundle voice with access and applications.

- *New protocols lower the barriers to applications development.* Session initiation protocol (SIP) is a proposed standard that distributes call processing away from the switch to software in end-user devices. In addition, the protocol allows end users to customize applications; for example, to create custom voice mail menus or add a "click-to-talk" link on a Web site. Microsoft will include a SIP stack in its new Windows XP operating system, which will allow PCs to make phone calls through a VoIP service provider like Net2Phone.

- *Software improves VoIP quality.* Startups like Kagoor Networks have developed routing and call-management software that prioritizes voice packets over other content, thus cutting down on the latency and jitter that degrade quality. iBasis, a VoIP carrier, uses its own proprietary routing capability to send 95 percent of its calls over the public Internet while maintaining toll quality. Continued cost savings will come from compression improvements and the ability to carry multiple calls on the same line.

Further, Forrester says that the distributed and open nature of packet voice technologies enables telephony providers to create new applications *and* save money. For example, carriers that implement softswitches and distribute SIP-based devices will be able to:

- *Offer simpler, better Centrex.* Companies like GoBeam and Ascentel Computing offer network-based services that support voice applications like automated call routing and desktop conferencing. These firms target customers that want to outsource their voice systems to avoid expensive equipment purchases but retain sophisticated custom-calling features and the ability to self-provision changes via the Web (see the June 2001 Forrester Report "IP Phones: Better, Not Cheaper").

- *Turn the phone into a ubiquitous Web client.* Voice portals like Tellme Networks and BeVocal enable access to information, like driving directions and weather reports, through a verbal interface. These companies unlock the information and transaction capabilities of the Internet to the everyday telephone and increasingly prevalent mobile phone. AOL By Phone now lets members use any touch-tone phone to check e-mail and stock quotes.

· *Bundle new services.* Rather than buy Nortel or Lucent switches, companies like Broadview Networks install less costly gear to lower startup costs. Why sell voice in the first place? In addition to new revenue streams, it reduces churn rates. Cox Communications, which sells video, data, and voice services, claims that up-selling customers from one service to all three reduces churn rates by 60 percent.

All categories of service providers, whether established, start-up, or next generation, have embarked on the road to IP telephony. Table 10.15 shows the various packet voice activities that service providers are currently engaged in.

TABLE 10.15   Service Providers' Packet Voice Activities (Source: Forrester Research, Inc.)

SERVICE PROVIDER	PLAYERS	WHAT THEY'RE DOING	COMMENTS
Long distance	AT&T Consumer Services, MCI WorldCom, Qwest, Sprint	• WorldCom offers VoIP on dedicated connections for businesses • Qwest wholesales IP network access for voice and data	• ILEC entry into LD will drive prices lower • VoIP equipment tested regularly
Local	BellSouth, Qwest, SBC, Verizon	• Entering new LD markets • Verizon uses VoIP gateways for international traffic	• Focused on entering LD • Limited packet voice tests in progress
Next-generation providers	Dialpad, Genuity, iBasis, ITXC, Net2Phone, Level 3	• VoIP over broadband; PC to phone; PC to PC; VoIP calling cards; international VoIP transport • ITXC offers "click-to-talk" services for call centers	• New data providers hope voice will drive capacity utilization • VoIP players will resell services through carriers
Cable	AT&T Broadband, Cox, Comcast, Time Warner Cable	• AT&T and Cox sell voice today • Time Warner Cable is test-marketing VoIP as part of its broadband service	• Encroach on local phone monopoly by selling voice • Circuit-switched efforts have slowed in anticipation of more mature VoIP solutions

BIBLIOGRAPHY

Allegiance Research Report (source: RHK), August 2001.

Business-Class Service at a Coach Price? Claudia Bacco, tele.com, September 5, 2001.

Christine Heckart, TeleChoice, August 2001.

CLECs: A Call to Arms. Ted McKenna, staff editor, *Telecommunications Magazine*, April 2001 (online).

Eight Lessons from the Telecom Mess. Commentary. Steve Rosenbush and Peter Elstrom. *Business Week*, August 13, 2001.

Hear No Risk, See No Risk, Speak No Risk: How a Bunch of Wall Street Analysts and Others Hyped a Company called Winstar—to death. Bethany McLean. *Fortune Magazine*, May 14, 2001.

Interview with CTO and cofounder Rashmi Doshi of Everest Broadband Networks, April 2001.

Interview with Chris Heckart. TeleChoice, August, 2001.

Interview with Jim Marsh, The Management Network Group (TMNG), June 2001.

Interview with Katie Wacker, director of corporate communications at McLeod USA, August 2001.

Kennard Sticks Up for Small Telcos. Richard Martin and Aaron Pressman, *The Industry Standard*, May 23, 2001.

Marketing Voice over DSL to Business Users, TeleChoice, November 2000.

Successful Niche Marketing. Gerry Blackwell ClecPlanet.

Telecom Debt Debacle Could Lead to Losses of Historic Proportions. Gregory Zuckerman and Deborah Solomon, (Telecommunications) Staff Reporters, *The Wall Street Journal*, May 11, 2001).

Telecom Meltdown, Peter Elstrom, *Business Week*, April 23, 2001.

Telecom's Wake-Up Call. Peter Elstrom, with Michael J. Mandel (New York). *Business Week*, September 25, 2000.

The One CLEC You Should Consider. Roben Farzad, SmartMoney.com, July 18, 2001.

"To Build or Not to Build: The Economics of a Network," D. Gregory Smith of Z-Tel Technologies, April 4, 2001.

INDEX

Note: Boldface numbers indicate illustrations.

ABOUT THE AUTHOR

Rajoo Nagar is President of PowerTel Consulting of Cupertino, California, which provides consulting and marketing services to emerging telecommunications companies. Prior to founding PTC, she served in senior marketing roles at TollBridge Technologies, Tut Systems, and PowerTel Global, which she co-founded. Ms. Nagar holds master's degrees in computer science and business management, and was a planning engineer at Hewlett-Packard prior to entering the world of telecommunications.